Control Software for Mechanical Systems

Object-Oriented Design in a Real-Time World

IADT LIBRARY
915 NATIONAL PARKWAY
SCHAUMBURG, IL 60173

ISBN 0-13-786302-0

Control Software for Mechanical Systems:
Object-Oriented Design in a Real-Time World

D. M. Auslander

J. R. Ridgely

J. D. Ringgenberg

Prentice Hall PTR
Upper Saddle River, New Jersey 07458
www.phptr.com

A Cataloging-in-Publication Data recond for this book can be obtained from the Library of Congress.

Editorial/Production Supervision: Mary Sudul
Acquisitions Editor: Bernard Goodwin
Editorial Assistant: Michelle Vincente
Marketing Manager: Dan DePasquale
Manufacturing Manager: Alexis Heydt-Long
Cover Design Director: Jerry Votta
Cover Designer: DesignSource
Compositor: Lori Hughes

©2002 Prentice Hall PTR
A division of Pearson Education, Inc.
Upper Saddle River, NJ 07458

Prentice Hall books are widely used by corporations
and government agencies for training, marketing, and resale.

For information regarding corporate and government bulk discounts, contact:

Corporate and Government Sales: (800) 382-3419 or corpsales@pearsontechgroup.com

All products mentioned herein are trademarks or registered trademarks of their respective owners.

All rights reserved. No part of this book may be reproduced, in any form or by any means, without permission in writing from the publisher.

Printed in the United States of America
10   9   8   7   6   5   4   3   2   1

ISBN 0-13-786302-0

Pearson Education LTD.
Pearson Education Australia PTY., Limited
Pearson Education Singapore, Pte. Ltd.
Pearson Education North Asia Ltd.
Pearson Education Canada, Ltd.
Pearson Educación de Mexico, S.A. de C.V.
Pearson Education—Japan
Pearson Education Malaysia, Pte. Ltd.

Contents

PREFACE	xiii
1 MECHATRONICS	**1**
1.1 A History of Increasing Complexity	2
1.2 Mechatronic System Organization	3
1.3 Amplifiers and Isolation	3
1.4 Scope: The Unit Machine	4
1.5 Control	5
1.6 Real-Time Software	5
1.7 Nasty Software Properties	8
1.8 Engineering Design and Computational Performance	9
1.9 Control System Organization	10
1.10 Software Portability	10
1.11 Operator Interface	11
1.12 Multicomputer Systems: Communication	12
1.13 The Design and Implementation Process	12
1.13.1 Performance Specification	12
1.13.2 Design Documentation	13
1.13.3 Simulation	14
1.13.4 Laboratory Prototype	15
1.13.5 Production Prototype	16
1.13.6 Production System	17
1.13.7 Maintenance	17
2 TASKS	**19**
2.1 Example: Task Selection in a Process System	20
2.2 Tasks and the Control Hierarchy	21
2.2.1 Intertask Communication	22
2.3 Task Structure Examples	22
2.3.1 Velocity Control of a DC Motor	23
2.3.2 Heater Control	24
2.3.3 Toaster Oven	26
2.3.4 Flexible Position Control of a DC Motor	27

	2.4 Simulation	29
	2.5 More Task Structure Examples	31
	2.5.1 Coordinated, Two-Axis Motion	31
	2.5.2 A Washing Machine	32
3	**STATE TRANSITION LOGIC**	**35**
	3.1 States and Transitions	36
	3.2 Transition Logic Diagrams	36
	3.3 Tabular Form for Transition Logic	37
	3.4 Example: Pulse-Width Modulation (PWM)	38
	3.5 Transition Logic for the Process Control Example	39
	3.6 Nonblocking State Code	41
	3.7 State-Related Code	41
	3.8 State Scanning: The Execution Cycle	42
	3.9 Task Concurrency: Universal Real-Time Solution	43
4	**DIRECT REALIZATION OF SYSTEM CONTROL SOFTWARE**	**45**
	4.1 Language	45
	4.2 Time	47
	4.3 Program Format	47
	4.4 Simulation	48
	4.5 Simulation in Matlab	48
	4.5.1 Templates for Simulation Using Matlab	48
	4.5.2 Simulation of PWM Generator	53
	4.5.3 Simulation of Three-Tank Process System	57
	4.6 Intertask Communication	61
	4.7 Real-Time Realization	62
	4.8 Real-Time Realization with Matlab	62
	4.8.1 Heater Control Implementation in Matlab	63
5	**SOFTWARE REALIZATION IN C++**	**67**
	5.1 Simulation in C++	67
	5.2 Templates for Simulation in C++ (group-priority)	68
	5.3 PWM Simulation Using C++ (group-priority)	80
	5.4 Simulation in C++ (with TranRun4)	82
	5.4.1 Components	82
	5.4.2 The Master Scheduler	84
	5.4.3 Process Objects and Task Lists	85
	5.4.4 Task Objects	86
	5.4.5 Tasks with No State Object	89
	5.4.6 Creating Task Classes	89
	5.4.7 State Objects	91
	5.4.8 Creating State Classes	93
	5.4.9 The Main File and UserMain() Function	94
	5.5 Real-Time Realization with C++	97

6	**INTERTASK COMMUNICATION**	**99**
	6.1 Communication Within a Process	100
	6.1.1 Data Integrity	100
	6.1.2 Design Rules	102
	6.2 Communication Across Processes	105
	6.2.1 Message Passing	105
	6.2.2 Message Passing in the Group Priority Scheduler	106
	6.2.3 Message Passing in the TranRun4 Scheduler	112
	6.2.4 Distributed Database	115
	6.2.5 Distributed Database in the Group Priority Scheduler	116
	6.2.6 Distributed Database in the TranRun4 Scheduler	118
7	**TIMING TECHNIQUES ON PC COMPATIBLES**	**121**
	7.1 Calibrated Time	121
	7.2 Free-Running Timer	122
	7.2.1 Hardware Timers on the PC	123
	7.2.2 Performance Timers in Unix and Windows	124
	7.3 Interrupt-Based Timing	125
8	**MULTITASKING: PERFORMANCE IN THE REAL WORLD**	**127**
	8.1 Priority-Based Scheduling—Resource Shifting	127
	8.1.1 Continuous vs. Intermittent Tasks	128
	8.1.2 Cooperative Multitasking Modes	129
	8.2 Matlab Template for Minimum-Latency Dispatcher	131
	8.2.1 Example: Simulation of PWM-Actuated Heater	131
	8.3 Cooperative Multitasking Using C++	133
	8.3.1 Inheriting Task Behavior—Two PWMs	137
	8.4 Preemptive Multitasking Modes	138
	8.5 Realization of Interrupt-Based Dispatching	140
	8.5.1 How Many Priority Levels Are Necessary?	141
	8.5.2 Which Interrupt Sources Will Be Used?	141
	8.5.3 Interrupt-Based Dispatching Functions	142
	8.5.4 Attaching Dispatching Functions to Interrupts	143
9	**A CHARACTER-BASED OPERATOR INTERFACE**	**145**
	9.1 Operator Interface Requirements	145
	9.2 Context Sensitive Interfaces	146
	9.3 User Interface Programming Paradigms	147
	9.4 Mechatronics System Operator Interface	147
	9.5 Operator Interface Programming	148
	9.5.1 The Operator Screen	148
	9.5.2 Programming Conventions in C++	149
	9.5.3 Heater Control Operator Interface	151

10 GRAPHICAL OPERATOR INTERFACES — 155
- 10.1 Graphical Environments — 156
  - 10.1.1 Windowing Software: Events and Messages — 156
  - 10.1.2 Operator Interface vs. Standard Windowing Application — 157
  - 10.1.3 Simplified Programming for Windowing Systems — 157
  - 10.1.4 The Ease-of-Use Challenge — 158
  - 10.1.5 Methods of Simplifying Window-Style Programming — 158
- 10.2 The Times-2 Problem — 158
  - 10.2.1 Times-2: Character-Based Interface — 158
  - 10.2.2 Times-2: Visual Basic — 160
  - 10.2.3 Times-2: Bridgeview — 162
- 10.3 Screen Change — 166
  - 10.3.1 Screen Change in Visual Basic — 166
  - 10.3.2 Screen Change: Bridgeview — 168
- 10.4 Heat Exchanger Control in Bridgeview — 171
- 10.5 Interprocess Communication: DDE — 173
  - 10.5.1 DDE: The C++ Side — 173
  - 10.5.2 Communicating with Excel — 176
  - 10.5.3 A DDE Server in C++ — 177
  - 10.5.4 DDE Communication Between C++ and Visual Basic — 178
  - 10.5.5 DDE Communication Between C++ and Bridgeview — 180
- 10.6 Putting It All Together — 181

11 DISTRIBUTED CONTROL I: NET BASICS — 185
- 11.1 Multiprocessor Architectures — 186
  - 11.1.1 Symmetric Multiprocessing (SMP) — 186
  - 11.1.2 Buses — 186
  - 11.1.3 Networks — 187
  - 11.1.4 Point-to-Point Connections — 189
- 11.2 TCP/IP Networking — 190
  - 11.2.1 The Physical Context — 190
  - 11.2.2 Interconnection Protocols — 191
  - 11.2.3 TCP and UDP — 191
  - 11.2.4 Client/Server Architecture — 192
- 11.3 Implementation of UDP — 192
  - 11.3.1 Sockets — 192
  - 11.3.2 Setting Up for Network Data Exchange — 193
  - 11.3.3 Nonblocking Network Calls — 195
  - 11.3.4 Receiving Information — 195
  - 11.3.5 Client-Side Setup — 196
- 11.4 The Application Layer — 197
  - 11.4.1 Data Coding — 197
  - 11.4.2 Building the Packet — 198
  - 11.4.3 Parsing a Packet — 200

12 DISTRIBUTED CONTROL II: A MECHATRONICS CONTROL APPLICATION LAYER — 203
- 12.1 Control System Application Protocol — 203
- 12.2 Startup of Distributed Control Systems — 207
- 12.3 Testing the Application Protocol — 208
- 12.4 Using the Control Application Protocol — 209
- 12.5 Compiling — 212

13 JAVA FOR CONTROL SYSTEM SOFTWARE — 213
- 13.1 The Java Language and API — 214
  - 13.1.1 Networking — 214
  - 13.1.2 AWT/Swing — 214
  - 13.1.3 Multithreading — 214
- 13.2 Preconditions for Real-Time Programming in Java — 215
  - 13.2.1 Deterministic Garbage Collection — 215
  - 13.2.2 Memory and Hardware Access — 215
  - 13.2.3 Timing — 216
- 13.3 Advantages of Java for Control Software Design — 216
  - 13.3.1 Modularity — 216
  - 13.3.2 Distributed Control — 217
  - 13.3.3 Platform Independence and Prototyping — 217
  - 13.3.4 Operator Interface Design — 217
- 13.4 Java and the Task/State Design Method — 218
  - 13.4.1 Inner Classes — 218
  - 13.4.2 Networking — 218
  - 13.4.3 Documentation — 219
- 13.5 The Current State of Real-Time Java — 219

14 PROGRAMMABLE LOGIC CONTROLLERS (PLCs) — 221
- 14.1 Introduction — 221
- 14.2 Goals — 222
- 14.3 PLC Programming — 223
  - 14.3.1 When to Use a PLC — 223
  - 14.3.2 Ladder Logic — 224
  - 14.3.3 Grafcet/Sequential Flow Charts — 226
- 14.4 The Task/State Model — 226
- 14.5 State Transition Logic for a PLC — 227
  - 14.5.1 State Variables — 227
  - 14.5.2 Ladder Organization — 227
  - 14.5.3 Transitions — 228
  - 14.5.4 Outputs — 229
  - 14.5.5 Entry Activity — 229
  - 14.5.6 Action Outputs — 229
  - 14.5.7 Exit (Transition-Based) Outputs — 229
  - 14.5.8 Common Exit Activities — 230

14.6	PLC Multitasking	230
14.7	Modular Design	231
14.8	Example: Model Railroad Control	231
14.9	Simulation – Portability	232

15 ILLUSTRATIVE EXAMPLE: ASSEMBLY SYSTEM — 235

- 15.1 The Assembly System — 235
- 15.2 System Simulation — 237
- 15.3 Development Sequence — 237
- 15.4 Belt Motion Simulation (Glue00) — 238
  - 15.4.1 Modeling Belt Dynamics — 238
  - 15.4.2 Definition of Task Classes — 239
  - 15.4.3 Instantiating Tasks: the Main File — 241
  - 15.4.4 The Simulation Task — 242
  - 15.4.5 The Data Logging Task — 244
  - 15.4.6 Timing Mode — 245
  - 15.4.7 Compiling — 246
  - 15.4.8 Results — 246
- 15.5 Oven Temperature Simulation (Glue01) — 247
- 15.6 PID Control of Belt Position and Oven Temperature (Glue02) — 247
  - 15.6.1 Keeping Classes Generic — 248
  - 15.6.2 The PIDControl Class — 248
  - 15.6.3 Results — 250
- 15.7 Better Control of Motion (Glue03) — 250
  - 15.7.1 Trapezoidal Motion Profile — 251
  - 15.7.2 Motion Profile Class — 252
  - 15.7.3 Profiler State Structure — 253
  - 15.7.4 Round-Off Error — 257
  - 15.7.5 Discretization Errors in Simulation — 257
- 15.8 A Command Structure for Profiled Motion (Glue04) — 260
  - 15.8.1 Message-Based Command Structure — 260
  - 15.8.2 State Transition Audit Trail — 261
  - 15.8.3 Motion Results — 263
- 15.9 Clamps (Glue05) — 263
- 15.10 Robots (Glue06) — 265
- 15.11 Cure/Unload (Glue07) — 266
- 15.12 Making Widgets (Glue08) — 271

16 THE GLUING CELL EXERCISE IN TRANRUN4 — 273

- 16.1 The Gluing System — 273
- 16.2 Simulation and Prototyping — 274
- 16.3 The Project Components — 274
- 16.4 Glue00: Conveyor Simulation — 275
  - 16.4.1 The Dynamic Model — 275
  - 16.4.2 Creating the Conveyor Task — 277

16.4.3 The Data Logging Task	280
16.4.4 Data Communication Between Tasks	284
16.4.5 The Main File	286
16.4.6 Glue00 Results	288
16.5 Glue01: An Oven Simulation	288
16.5.1 Configuration and Status Printouts	289
16.6 Glue02: PID Control	291
16.7 Glue03: The Operator Interface	292
16.7.1 Results	297
16.8 Glue04: Motion Profiling	299
16.9 Glue05: Belt Sequencing	306
16.10 Glue06: The Glue Application Machine	307
16.11 Glue07: Transport Task Supervision	309
16.12 Glue08: The Completed Assembly System	311

17 THE GLUING CELL EXERCISE IN TRANRUNJ — 315

17.1 Getting Started	315
17.1.1 Program Entry Point	315
17.1.2 The userMain Method	316
17.2 Writing Custom Tasks and States	317
17.2.1 Creating a Task Class	317
17.2.2 Creating a State Class	319
17.3 Implementing State Transition Logic	320
17.4 Global Data and Intertask Messaging	321
17.4.1 Global Data Items	321
17.4.2 Task Messages	322
17.5 Continuous vs. Intermittent Tasks	323
17.6 Scheduler Internals	324
17.6.1 Operating System Processes vs. CProcess	324
17.6.2 Foreground vs. Background Execution Lists	325
17.6.3 Scheduling Modes	325
17.7 Execution Profiling	325
17.8 Intertask Messaging Across Different Processes	326
17.9 Tips And Tricks	328
17.9.1 Judicious Use of Execution-Time Profiling	328
17.9.2 Integer Labels for Global Data and Task Message Inboxes	328
17.9.3 The TaskMessageListener Interface	328
17.9.4 Scheduler Sleeping	329
17.9.5 Anonymous State Classes	329
17.10 Additional Information	330

BIBLIOGRAPHY — 331

INDEX — 333

PREFACE

The control of complex mechanical systems often falls between the cracks of engineering curricula, but the topic occupies an extremely important place in the world of industrial control. Most courses and professional reference material cover the details of embedded control of very small systems using microcontrollers, as well as details of electronics and control theory. In contrast, this book addresses issues associated with the design of control software for mechanical systems consisting of significant numbers of sensors and actuators – systems whose activities must be coordinated. These systems often include a computer-based operator interface and internal as well as external network connections.

The term "mechanical system" in the context used here refers to a system in which real physical power must be delivered to a target object. The power could be in a form such as motion, heat, force, pressure, or many others, so the range of applicability is very large. The domain of complexity encompasses what we call a "unit machine," a mechanical system in which all parts either exchange physical power directly or exchange material with little or no buffering. This definition makes a distinction between the control of a specific unit machine or process on the one hand, and designing control software for work cells, entire process plants, or sections of a factory on the other.

The material we present was developed for teaching both undergraduate and graduate courses to classes consisting mostly of mechanical engineering students. Mechanical engineering students are often not experienced in computer programming and therefore have to learn an appropriate programming language (more on that in a moment), learn the design methodology, and develop the ability to apply both in a series of lab exercises and projects. While a 15 week semester makes for a somewhat compressed timetable to do all of this, the students seem to emerge with a good understanding of how the pieces fit together. It is not unusual for many students to progress directly to successful work in industrial applications based mostly on the exposure in this course. The material is usable in other configurations as well, including self study, multi-semester/quarter format with more emphasis on design or capstone projects, and so on.

The presentation in this text is based on a philosophy in which the control engineering insight is embedded in a design layer rather than in computer code. Thus, issues of portability, design review, and internal communication can be stressed. The generation of functioning computer code is a separate process where considerations of real time constraints, costs of computing hardware and software, and ease of maintenance can be considered. Maintaining this separation is crucial to the idea that production of control software for mechanical systems must be a predictable engineering process.

The methodology we use for design specification is based on finite state machines and semi-independent tasks. It has proved to be a methodology capable of handling the level of complexity needed for these systems. It has some common elements with more formal methodologies such as the universal modeling language (UML) but is considerably simpler. The choice of methodology is based on its ability to describe control behavior in a way that is easily explainable to a broad audience, as well as the complementary property of easy hand translation for computer implementation in almost any language or environment. Software portability to achieve maximum protection to software investment is emphasized throughout.

Choosing computer languages provokes as much vehement discussion as religion or politics! To neutralize some of that partisanship, we have kept the design layer as primary, with the actual software implementation kept as fluid as possible. Nonetheless, it is necessary to choose some language in which to implement actual control systems. While C is probably the most common language in use today, the structures of C++ and Java provide us with more assistance in building a reusable software infrastructure that makes the design model as obvious as possible. We have therefore used C++ and Java for most of the actual control implementation and for the example programs used throughout this text. Java has the advantage of greater portability and is a cleaner language than C++ because it does not need backward compatibility with C. Its syntax includes graphic user interface (GUI) constructs and TCP/IP networking, for example, which are not included in the C++ syntax and are thus not portable for C++ applications. Java's class files are executable on any platform supporting Java, so cross-development is greatly simplified. On the other hand, Java is considerably slower by virtue of its use of a virtual machine for execution; and features such as garbage collection are tricky to deal with in a high speed, real time environment.

The basic programming model, however, is easily adapted for other programming languages. Sample software for simulation using Matlab is given in the book and *ad hoc* implementations in C or other purely algorithmic languages are easily constructed.

We have three main software packages available to support development, two in C++ and one in Java. All of these support the basic task/state design model, and all can operate in a variety of environments to support simulation, prototyping, and even production versions of software. The two C++ packages differ in that one uses a much simpler scheduling module, so more of the decision-making about when tasks should execute is left to the programmer. The other C++ package, and the

Java package, support a richer scheduling module. The simpler C++ package and the Java package both support full TCP/IP networking in a manner that allows tasks to be distributed over multiple computers through very simple configuration changes, with no other changes needed in the control software.

The Windows NT family of operating systems (NT, 2000, and XP) have been used for a number of years as the main development environment. C++ and Java compilers are readily available as are cross-compilers for many target environments. They have also been used as real time execution environments for early prototyping and debugging. However, the timing in any of the Windows environments is not consistent to the sub-millisecond level needed for much mechanical system control, so other operating systems must be used for higher quality implementation (although it is surprising how much can actually be accomplished without leaving Windows). In the past several years we have used DOS, Windows NT modified with RTX (from VenturCom), and QNX as real time platforms.[1] We are currently working with Java and using QNX as the real time platform, since neither DOS nor NT/RTX supports real-time Java effectively.

The operator interface is a critical part of many commercial control systems. Not only does the interface contribute to efficient use of the target system, but it often affects safety in critical ways. While a full discussion of human factors and other disciplines affecting operator interface design is beyond the scope of this book, we do discuss several means of constructing computer-based operator interfaces. The book shows samples based on using Bridgeview (an industrial relative of Labview, from National Instruments) and Visual Basic. These and other simplified GUI construction methods are important from an efficiency point of view in that they can often be used by control engineers to construct prototype operator interfaces without having to devote inordinate amounts of time to the process. The construction of GUIs in Java is not specifically discussed in the book, but available sample Java code demonstrates the use of Java's Swing components in rapid GUI development.

A detailed case study is provided to illustrate the use of the design and implementation methodology on a problem with reasonable complexity. The implementation is entirely a simulation, so it can be run anywhere Java or C++ can be installed – and thus it makes a good vehicle for use by a class or as a self-study example. Implementations are given for all three of the software infrastructures we have developed.

All of the software packages and sample software supporting the book are available from the web site of one of the authors:

www.me.berkeley.edu/~dma

Additional Java information is available from the web site of another author:

www.ugcs.caltech.edu/~joeringg/TranRunJ/

---

[1] DOS is not technically a real time platform, but it does not interfere with the real time constructs we have created and thus allows much higher performance real-time scheduling than Windows.

There is considerable teaching material available on the class web sites for the courses based on this book:

www.me.berkeley.edu/ME135 and www.me.berkeley.edu/ME230

This includes a current syllabus for each of these courses, lab assignments, notes, and other relevant information. There is also a large amount of class presentation material in PowerPoint format available; this can be requested by sending an email to dma@me.berkeley.edu.

Chapter 1

MECHATRONICS

Mechanical system control is undergoing a revolution in which the primary determinant of system function is becoming the control software. This revolution is enabled by developments occurring in electronic and computer technology. The term *mechatronics*, attributed to Yasakawa Electric in the early 1970s, was coined to describe the new kind of mechanical system that could be created when electronics took on the decision-making function formerly performed by mechanical components. The phenomenal improvement in cost/performance of computers since that time has led to a shift from electronics to software as the primary decision-making medium. With that in mind, and with the understanding that decision-making media are likely to change again, the following definition broadens the concept of mechatronics while keeping the original spirit of the term:

> **Mechatronics:** *The application of complex decision-making to the operation of physical systems.*

With this context, the compucentric nature of modern mechanical system design becomes clearer. Computational capabilities and limitations must be considered at all stages of the design and implementation process. In particular, the effectiveness of the final production system will depend very heavily on the quality of the real-time software that controls the machine.

By the mid-1990s, computers embedded in devices of all sorts had become so ubiquitous that the newspaper *USA Today* thought it important enough to show the following on the first page of the business section:

USA Today, May 1, 1995, page 1B
USA SNAPSHOTS - A look at statistics that shape your finances
Micromanaging our Lives
Microcontrollers – tiny computer chips – are now running things from microwave ovens (one) to cars (up to 10) to jets (up to 1,000).
How many microcontrollers we encounter daily:
    1985   less than 3
    1990   10
    1995   50

By 2000, "up to 10" microcontrollers in high-end cars had grown to 60[1]. In the same vein, according to *Embedded Systems Programming*: "More than 99% of all microprocessors sold in 1998 were used in embedded systems"[2].

1.1 A History of Increasing Complexity

Technological history shows that there has always been a striving for complexity in mechanical systems. The list in table 1.1 (which is not strictly chronological) contains some examples of machines that show the marvelous ingenuity of their inventors.

A diagram of the Watt steam engine speed control (governor) is shown in figure 1-1. As the engine speeds up, the flyballs spin outward, moving the linkage so as to close the steam valve and thus slow the engine. The entire control process is performed by mechanical hardware.

Table 1.1. Some Historically Significant Machines

Watt governor
Jacquard loom
Typewriter
Mechanical adding machine
Pneumatic process controller
Sewing machine
Tracer lathe
Cam grinder
Tape controlled NC machine tool

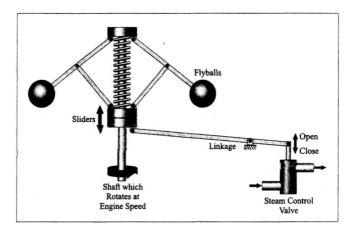

Figure 1-1. Watt governor

1.2 Mechatronic System Organization

Developments in electronics have made it possible to energetically isolate the four components making up a controlled mechanical system (figure 1-2).

Once isolated from the instruments on one side and the actuators on the other, computation can be implemented using the most effective computing medium available. There is no need to pass a significant amount of power through the computational element. The computing medium is now the digital computer, and the medium of expression for digital computers is software.

This ability to isolate is recent. Watt's speed governor, for example, combined the instrument, computation, and actuation into the flyball element. Its computational capability was severely limited by the necessity that it pass power from the power shaft back to the steam valve. Other examples in which computation is tied to measurement and actuation include automotive carburetors, mastered cam grinders, tracer lathes, DC motor commutation, timing belts used in a variety of machines to coordinate motion, rapid transit car couplings (which are only present to maintain distance; cars are individually powered), and myriad machines that use linkages, gears, cams, and so on to produce desired motions. Many such systems are being redesigned with software-based controllers to maintain these relationships, with associated improvements in performance, productivity (due to much shorter times needed to change products), and reliability. Some, such as transit car couplings, retain traditional configurations for a variety of reasons including cost, a lack of need for the added flexibility, and customer acceptance problems.

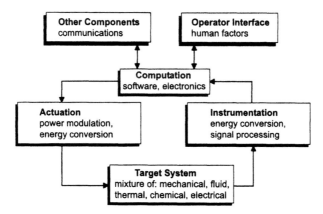

Figure 1-2. Parts of the system

1.3 Amplifiers and Isolation

The development of electronic amplification techniques has opened the door to mechatronics. Operational amplifier (op-amp) based circuits, such as those shown

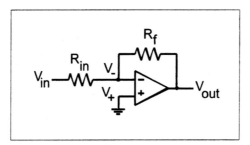

Figure 1-3. Inverting operational amplifier

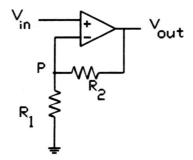

Figure 1-4. Non-inverting amplifier

in figures 1-3 and 1-4, can provide isolation and generation of linear and nonlinear, static and dynamic computing functions.

Developments in power electronics have provided the same kind of flexibility and isolation on the actuation side that op-amps provide on the signal side, and give a means for translating commands computed in electronic form into action.

Digital electronics has added enormously to the range of computational functions that can be realized. Digital electronics also form the basic technology for implementing computers, but with the use of computers it is software rather than electronics that represents the medium of expression. Although mechatronics based on electronic computing elements opened a new world of mechanical system control, software has extended that world in previously unimaginable ways.

1.4 Scope: The Unit Machine

Control systems can range from extremely simple to enormously complex. The same design methods and software development techniques will not work across this very wide range. This material is targeted to a middle level in the complexity spectrum: what we call the unit machine or unit system. In general terms, a unit machine encompasses a closely interacting set of physical components. Because of the close interaction there are tight time constraints that apply to the interactions between parts of the machine or system. More specifically, the parts of a unit machine either exchange real physical power, or exchange material with little or no buffering. The complexity can range from a few motors or heaters to a system with tens of such components. A unit machine can stand alone, or be part of a larger system such as a manufacturing cell or plant. However, when it is part of a larger system the interactions with other parts of the system are at a much slower time scale than the interactions within the unit machine.

The first step in any design process is to assess the scope of the problem. Underestimating the complexity is the most dangerous and most common error. If design methods are used that cannot handle the level of complexity needed, the chances of achieving a reliable, high performance control system are severely compromised. With luck, satisfactory performance may be achieved, but only with a loss in predictability of the design and implementation process. At worst, the system will simply never perform satisfactorily and will cost far more than initially estimated as well.

Overestimating complexity is not as dangerous, but it can lead to a loss of an economically competitive position. A system designed using methods suitable for much more complex systems will work well but may use more resources than would be necessary if a simpler method were used.

1.5 Control

The term *control* is used here in a broad context: that is, it encompasses all of the software, electronics, instruments, and actuators needed to achieve the goals of the target system. Our focus is specifically on the software, but control system software cannot be considered outside the context of the other elements of a control system. Control is often used to refer specifically to feedback control. Feedback control often, if not usually, present in mechatronic control systems, is only one small part of the overall control system. Thus, when used alone, control will refer to the broader context; feedback control will be identified explicitly when referring to the mathematics of feedback control or to implementations of feedback control algorithms within the broader control system.

1.6 Real-Time Software

The dominant implementation medium as well as primary medium of expression for contemporary mechatronic systems is real-time software. Real-time software is a subdomain of software generally with distinct requirements and characteristics. While implementation media are subject to change, software is so dominant now that any design methodologies that are used must allow for the effective implementation of control systems using real-time software.

Real-time software differs from conventional software in that its results must not only be numerically and logically correct, they must also be delivered at the correct time. A design corollary following from this definition is that real-time software must embody the concept of *duration*, which is not part of conventional software. Duration is such an important concept that it will be the focus of the documentation technique used as the heart of a design methodology for producing reliable control system software.

Two terms in common use in this field have overlapping, but not identical meanings. These are real time, as used here, and embedded. Embedded refers to computers that are parts of other items and are not explicitly visible to the user. For example, a mobile (cell) phone has one or more embedded computers in it. These computers control the user interaction via the keyboard and display, control the connection between the phone and its nearest base station, filter the incoming signals to remove noise, encrypt and decrypt messages for security purposes, and so on. On the other hand, a pocket size personal digital assistant (PDA) *is* a computer. Even though the computer in it is of roughly the same capability and performs many of the same functions as the computer in the mobile phone, it is not thought of as embedded. Embedded computers may—or may not—perform functions that require real-time software. The functions they perform may or may not have time critical components. Real-time software, however, can run on embedded computers or on computers which are doing control functions but are clearly free-standing and not built into the target system.

Thus, many, but not most, embedded computers run real-time software, and many, but not most, real-time software runs on embedded computers.

An example showing the need for on-time delivery of results is the masterless cam grinder shown in figure 1-5. In the masterless cam grinder, the cam which is being ground to shape is rotated while the linear positioner moves back and forth to generate the desired shape for the cam. The rotation of the cam must be precisely coordinated with transverse motion to cut the noncircular cam profile.

This is an example of how mechatronics can change the way that problems are approached. In a conventional, mastered cam grinder the motion is controlled by a mechanical system which traces the contour of a master cam. The mechanical tracing mechanism must fill the role of computation (transmission of the desired contour) but also carry enough power to operate the mechanism. This is a limit to performance and can cause a considerable loss of productivity since the time to change the machine from one cam profile to another is considerable. Performance limitations come about because of the dynamics associated with the accelerations of the elements of the mechanism carrying the information from the master to the blank cam, because of friction in that mechanism, and because of other real-world effects. The result is that for high-precision cams, such as those used for automotive engines, the master usually has to be made slightly differently than the actual desired shape of the final product to compensate for these changes. Ironically, given the automotive application, the question may be asked: Why use a cam at all to manipulate the engine valves? In fact, there has been considerable work on electrically actuated valves, but they have not yet been used on mainline production automobiles.

Section 1.6. Real-Time Software

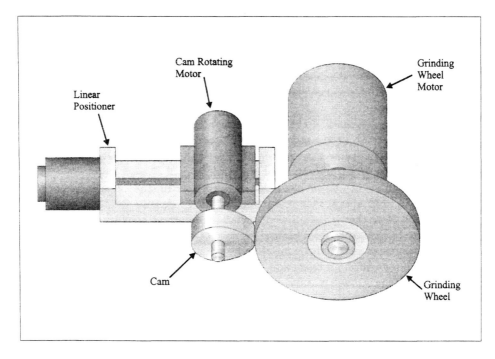

Figure 1-5. Masterless cam grinder

In the mechatronics-based cam grinder, the profile is stored in an array of numbers in software, so the machine's performance is limited only by its inherent mechanical properties. Changing the machinery to make a different cam profile is also much faster since it only requires the loading of new data instead of the installation of new mechanical hardware in the form of the master cam.

The real-time software used in most mechanical system control is also safety critical. Software malfunctions can result in serious injury and/or significant property damage. In discussing a series of software-related accidents which resulted in deaths and serious injuries from clinical use of a radiation therapy machine called the Therac-25, Leveson and Turner[3] established a set of software design principles "...that apparently were violated with the Therac-25..." These principles are:

- Documentation should not be an afterthought.

- Software quality assurance practices and standards should be established.

- Designs should be kept simple.

- Ways to get information about errors—for example, software audit trails—should be designed into the software from the beginning.

- The software should be subjected to extensive testing and formal analysis at the module and software level; system testing alone is not adequate.

In particular, it was determined that a number of these problems were associated with asynchronous operations which, while uncommon in conventional software, are the heart and soul of real-time software. Asynchronous operations arise from the interaction of the software with the physical system under control. In most cases, the physical system is made up of many components, all operating simultaneously. Although the function of the software is to coordinate this operation, events occurring at nearly the same time in the physical system do not necessarily have a predictable order. Thus, operation of the system may change from one cycle to another if the order of one or more of these events changes. The software must be designed in such a way that it operates correctly despite this lack of detailed predictability. Note that if the order of events is important to correct system operation, the control software must assure the correct order. However, there are many events that occur at nearly the same time for which order is not important.

Because of asynchronous operation, it becomes impossible to predict when groups of program statements will execute relative to other groups, as it is in synchronous software. Errors that depend on a particular execution order will only show up on a statistical basis, not predictably. Thus, the technique of repeating execution until the source of an error is found, which is so effective for synchronous (conventional) software, will not work for this class of real-time error.

In a like manner, there are errors that depend on the coincidence of certain sections of program code and events taking place in the physical system. Again, because these systems are not strictly synchronized, they can only have statistical characterization.

1.7 Nasty Software Properties

Software complexity tends to increase exponentially with size. What starts out as a small, manageable project can expand into something which literally never gets finished. To compound this problem, the economics of software-based products are such that over the lifetime of the product, maintenance costs dominate the total expense.

The complexity leads to a severe lack of predictability in software projects: "A 1997 survey indicated that 80% of all embedded software projects fail to come in on time"[4]. "No surprises" is the hallmark of effective engineering. While it is all too easy to treat programming as an unstructured activity, engineering reality demands a higher standard so that, to the extent possible, software time, cost, and size are reasonably predictable.

Both the complexity and the asynchronous nature of real-time software lead to situations in which bugs can remain dormant for long time periods, waiting to do their dirty work at the most inopportune time.

The antidotes to these problems flow from adopting an engineering perspective such that code writing (programming) should *not* be viewed as a creative activity!

On the engineering side, a method must be found to describe in detail how the control software should behave (a *specification*). This enables the needed engineering review function and enables the act of code production to become much more routine since the creative engineering part has been done at the descriptive level.

The second way of avoiding at least some of the complexity problems is to modularize the software and its production process. In this way, specific pieces of software are directly connected to certain machine functionality. System malfunctions can thus be traced to the relevant software more quickly, system changes can be made with reasonable assurances that the changes will not adversely affect some unexpected part of the machine, and individual software elements can be tested more easily. Modularization is also important in achieving reasonable delivery times for large projects. The modules can be worked on simultaneously by different people (perhaps in different parts of the world) and integrated into the final control system as they are produced.

1.8 Engineering Design and Computational Performance

Too often, the only detailed documentation of real-time software is the program itself. Furthermore, the program is usually designed and written for a specific real-time target environment. The unfortunate consequence of this is that the engineering aspects of the design problem become inextricably intertwined with the computation aspects. This situation relates directly to the first design principle listed earlier, "...documentation should not be an afterthought." If the system engineering is separated from the computational technology, the documentation will have to exist independent of the program; otherwise, documentation can be viewed as an additional burden.

The following definitions will be used to separate these roles as they appear in the proposed methodology:

> **System engineering:** Detailed specification of the relationship between the control software and the mechanical system.
>
> **Computational technology:** A combination of computational hardware and system software that enables application software based on the engineering specification to meet its operational specifications.

Using these definitions, the engineering specification describes how the system works; the computational specification determines its performance. As a result of this separation, if a suitable paradigm is adopted to describe the engineering specification, a much broader discussion and examination can be brought to bear because the details of the engineering can be discussed by project participants familiar with the problem, not just those familiar with computing languages and real-time programming conventions. This provides for meaningful design review of projects that are software intensive.

1.9 Control System Organization

The division into system engineering and computational technology allows separation of these two aspects of a control system. However, still further modularization is needed to meet the need for a rational design procedure. A two-level organization is used on both the engineering and computational sides of the control software design.

On the engineering side, a job is organized into

- **tasks**
- **states**

Tasks represent semi-independent activities of the control system. The division of the control software into tasks is the first level of modularization. Because the activities the tasks control are semi-independent, the tasks must all be active (execute) at the same time.

The next level of modularization is to states. States describe specific, sequential activities within tasks. While a task, for example, might control a specific motor, the states within that task could correspond to the motor turned off, the motor operating at constant velocity, the motor decelerating to a stop, and so forth.

The breakdown of the engineering specification into tasks and then states is a subjective process, requiring considerable engineering judgment. This stage is the primary creative design step.

On the computational side, the computing process is organized into

- **processes**
- **threads**

This breakdown is done to utilize the computing resources as efficiently as possible. It is independent of the task-state design in the sense that many different process-thread configurations can be devised for exactly the same task-state specification. In the absence of any economic constraints it is normally possible to implement a control system design with a single process and single thread, while still meeting all real-time specifications. However, economics always plays a role in engineering design, so effective process-thread design is crucial to commercially successful projects.

1.10 Software Portability

In mechanical system control, portability has consequences both for product lifetime and for the design/development cycle. The mechanical part of the system can have a commercial lifetime of anywhere from 5 to 20 years. On the other hand, the computational technology used for its control has a lifetime of only 3 to 5 years. To remain competitive, new versions of the product need to use new computers to take advantage of ever-increasing computational capability. Doing this effectively requires software that can easily be "ported" from the original target processor to new ones.

The need for software portability seriously affects the design/implementation cycle as well. Early stages of the software tend to be simulations, done to test hypotheses and to substitute for hardware not yet built. Later stages use laboratory prototypes, then pilot prototypes, then, finally, the actual production system. If software can't easily be migrated from each step to the next in this process, the whole process can become greatly elongated as new software must be created for each step, and there are significant possibilities for introducing new bugs at each stage.

Portability is complicated by real-time constraints. If real-time software environments (such as real-time kernels, schedulers, and operating systems) are used as aids in meeting those constraints, software written for one environment can require substantial rewriting to run in another. Crossing the full spectrum from simulation to production traverses environments in which program structure itself is incompatible. The methodology proposed in this book provides a layer of abstraction one level higher than the real-time environments and thus offers a means of bypassing these incompatibility problems.

1.11 Operator Interface

Many mechatronic systems require interaction with a human operator. The interaction can range from an ON/OFF switch to a fully-animated, multiscreen display. Other systems don't require any operator interface at all because they are themselves controlled by computer systems located in other parts of a larger system. Many of the computers in an automobile, for example, have no explicit operator interface and are buried deep within engine controls, braking systems, and so on. Some of them, however, do use an operator interface — in this case, the highly specialized operator interface used by the driver. The accelerator pedal serves as a command for power, the brake pedal for deceleration. Both are currently hybrid controls. They operate physical components directly and also give command signals to control computers. As cars become even more computer controlled, most of the direct connections will probably disappear. On many airplanes, they already have disappeared. Most large commercial and military aircraft in production at the turn of the century have "fly-by-wire" control systems in which the pilot's controls are linked only to computers — there is no direct physical link between the pilot and the control surfaces.

The automobile and airplane examples point out the most important property of operator interfaces: done well they facilitate proper system operation; done poorly they can lead to severe safety problems, particularly during unusual operating modes.

Operator interfaces, even those for fully computer-controlled systems, can be based on unique physical devices (such as the gas pedal on a car) or can use a computer-driven display. The design of the interface is a highly specialized activity and can consume a large portion of the implementation resources for a computer control system. Psychology, human-factors engineering, cultural factors, and various other considerations all play a part. Computer-based graphic displays cause particular problems for software portability because GUI (graphical user interface) software tends to be highly nonportable (see chapter 13 on Java for an exception to this).

1.12 Multicomputer Systems: Communication

All but the simplest mechanical control systems use more than one computer. The computers can be organized in a hierarchical arrangement or a "flat" one, or in both. Hierarchical organization is common in manufacturing facilities or other large systems where coordination of individual machines by management information systems assure efficient system operation. Flat architectures are used where several computers cooperate in the control of a single machine (or unit physical system). In either case, design of real-time software for controlling a mechanical system will often require understanding of multicomputer architectures and how those architectures will affect control system performance. Properly designed communication becomes a critical element in delivering a reliable system that can be implemented predictably.

As with operator interfaces, portability is a serious problem in the design and implementation of multicomputer control systems. At the earliest simulation levels of the design process, the whole control system will probably run on a single computer. As subsequent stages are implemented, a variety of architectures are likely to be used. In each of these stages, any number of network technologies could be utilized for intercomputer communication. There might even be several network and point-to-point communication technologies existing simultaneously in the final, production system. Within all of this diversity, the economics of software development dictate that as little change as possible be made to the software in going from one configuration to another.

1.13 The Design and Implementation Process

In the best of circumstances, the design of the control system will be completely integrated with the design of the physical target system itself, a process sometimes called control-configured design. In this way, control system considerations can be built in from the start, resulting in systems that perform better and are more reliable. Probably more often, the design of the target system is complete, and perhaps even built, before the control system design is started.

1.13.1 Performance Specification

In either case, the control system design process starts with an assessment of the system complexity and a judgment as to whether its operational goals are attainable. Given an affirmative judgment, a detailed performance specification is produced.

This starts from the original design goals of the target product, but will add specifications on startup and shutdown, transient behavior, response to disturbances, detection and response to system faults, safety, operator functions, maintenance modes, and so on. For projects in which the target system and control systems are designed simultaneously, this process often reveals ways in which changes in the basic system design can make control much easier to apply. Issues such as the types of motors and other actuators used, means of coupling power delivery to the target for the power, and decisions on use of software-based coordination versus mechanical couplings all relate strongly to control system design decisions.

Approval of the specification by all concerned parties is a major design step. It affirms in detail how the system is expected to behave. This specification is different from the original specification in most cases because the original specification is concerned mostly with productivity issues—how many or how much will be made and what quality level will be maintained. The control system specification contains much more because it considers all modes of system operation, not just normal production mode.

This specification is the basis for the control system software design. It also is the basis for control system acceptance testing. Development of an acceptance test at this stage is crucial. Otherwise, a divergence of expectations can cause real frustration and disappointment as the project nears its delivery date. Changes in the specification or acceptance tests need to be negotiated and agreed on as they arise [5], [6], [7].

Concurrent control system and target system design shows its benefits as early as the development of the performance specification. With control engineering input at this stage, unnecessarily strict specification items can be ameliorated and areas where performance can be improved without significant cost increases can also be identified. Unnecessary strictness usually refers to situations in which the cost is very high to achieve small increments in performance.

With the specification in hand, the rest of the design and implementation process becomes a series of translation and test steps, with each step either getting closer to the final product or causing a reversal to the previous step because of unexpected problems that were found.

A warning about testing: *It's hard!* Testing your own creation to destruction and finding faults in your own design are very difficult. Human nature is against it and as the designer, you know too much! Independent testing or quality control groups do a better job, but even in that case, colleagues and friends are involved. That said, continuous "tough" testing will perform wonders toward creation of a reliable and safe product on a predictable timetable.

1.13.2 Design Documentation

The first step in this process is to develop a formal design document. While the specification must be complete enough to fully describe the system's operation and specify the acceptance testing procedures, it does not generally describe the detailed operation of the control system. A formal design document does this using some sort of modular organization, often combined with one or more diagrammatic notations to describe data flow, operational modes, and so forth. The formal design documentation describes the system engineering aspect of the control system, as discussed in section 1.9. In addition to the functional description, it also must specify tolerances for accuracy of calculations, timing of responses to events, precision of external signal conversions, and other such issues that will affect performance.

Design review represents the testing stage for the formal design documentation. Despite its detailed nature, the methodology used for the design documentation must be simple enough to be understood by a wide spectrum of concerned individuals, not just engineers and programmers. A comprehensive design review at this stage will expose a surprising number of misunderstandings and conflicting expectations as to how the system should behave. Reconciling these differences early is incalculably valuable!

1.13.3 Simulation

Simulation should be the next step—"should" is used advisedly, because the simulation step is often skipped. Although it is hard to pin down properly controlled studies, there is reasonably wide belief that including a simulation step in the design process is an overall time saver. There are two reasons for skipping a simulation stage. First, in order to simulate the control system operation, the target system must be simulated. Taking the time to simulate the target system can seem to be an unneeded delay in getting the job done. Second, there may not be a suitable portable environment for implementing a simulation. Thus, when the simulation is done, the control software will have to be translated into another implementation medium.

The first argument is false economy. Any problems uncovered in the friendly environment of simulation can be solved far more economically than in the real physical environment. Even one such problem can probably more than pay for the cost of developing the simulation. Because any variables within a simulation can be logged and examined (i.e., the simulated system can be exhaustively instrumented), and because the asynchronous nature of the real system is simulated in a strictly synchronous computing environment, performance issues, computational sensitivities, unduly tight synchronization tolerances, logical errors in the design, unexpected interactions, and so on, can be found with relative ease. Fixing them may also be relatively easy, or may require extensive discussion, return to the design documentation stage, or even renegotiation of the performance specification.

The second argument has more merit, but even the cost of rewriting the control software is probably outweighed by the advantages of doing a simulation. The two ends of this spectrum are when the simulation is complex enough to need a dedicated simulation environment which does not readily interface with the language of the control software and when the control system software will use a proprietary environment such as a real-time operating system or programmable logic controller

(PLC) that is not readily reproduced on a general-purpose computational system (PC or workstation). Avoiding this situation if at all possible is a reason why control software portability is a priority in picking implementation tools.

Simulation is particularly important when the control system is in concurrent development with the target of the control. Not only is the actual system not yet built and thus not available for software development use, but lessons learned in testing the simulation can be fed back to the target system design process.

Testing the simulation starts with validation of the simulation of the target system. The nature of the testing is dependent on the level of detail of the model. The general rule is to use no more detail in the model than is necessary to test the control software.[1] Much more detailed models will probably be used by the target system designers, but for different purposes. Only when the control system has been validated against this simple model should it be integrated with the more complex model for use in concurrent system design.

With the simulation validated, the simulated control system software is first tested on a module-by-module basis. Tests should follow the performance specifications as well as the design documentation and should include exceptional circumstances as well as normal operation. After module-level testing is complete, the module integration is started, one module or as few modules at a time as possible. Again, testing at each stage follows the same guidelines used for module testing.

1.13.4 Laboratory Prototype

"There's many a slip twixt the cup and the lip" (from Greek mythology[2]). As valuable as simulation is, it doesn't tell the whole story. The next stage of testing utilizes laboratory conditions and equipment to test what has been learned from simulation on real hardware. Laboratory computers provide an excellent platform for this stage because of easy programmability, powerful processors, and large numbers of readily available products that can be used for software development, testing, and visualization. Problems resulting from asynchronous behavior, noisy signals, external interference, grounding problems, and less than ideal behavior of the target hardware all are exposed during this process. Production system economics are not a major issue, although some estimation of the computing resources that will be needed for the production system can be determined.

---

[1] "Pluralitas non est ponenda sine neccesitat" or "plurality should not be posited without necessity." The words are those of the medieval English philosopher and Franciscan monk, William of Ockham (ca. 1285-1349), and are commonly referred to as "Occam's Razor."

[2] The full story is: Ancaeos, helmsman of the ship Argo, after the death of Tiphys, was told by a slave that he would never live to taste the wine of his vineyards. When a bottle made from his own grapes was set before him, he sent for the slave to laugh at his prognostications; but the slave made answer, "There's many a slip 'twixt the cup and the lip." At this instant a messenger came in, and told Ancos that a wild boar was laying his vineyard waste, whereupon he set down his cup, went out against the boar, and was killed in the encounter. From the web version of The Dictionary of Phrase and Fable by E. Cobham Brewer, from the new and enlarged edition of 1894, http://www.bibliomania.com/Reference/PhraseAndFable/data/45.html

Testing follows the same model as was used for the simulation. The wide variety of analytic and data visualization tools available should be exploited to the fullest for diagnostic purposes. Easy file creation, network connectivity, and other convenient tools allow control system engineers to draw on skills throughout the organization to solve the difficult problems.

1.13.5 Production Prototype

The choice of a production computational environment is dominated by economics. The economics is a trade-off among the cost per unit of the computing hardware, the amortized cost of software development, and the value of short time to market. The result of the tradeoff is dramatically affected by the number of units expected to be produced.

Very large production runs magnify the importance of computing hardware cost. Hardware cost is minimized by using computing hardware with the minimum amount of specialty processing (such as floating point math hardware) and by making sure the maximum proportion of CPU cycles are used productively with the minimum memory footprint. This is inimical to rapid software development. Optimizing software for full CPU usage and doing mathematics (for example, a PID (proportional-integral-derivative) controller or a filter) with integer arithmetic add substantially to the software development cost.

Time to market is the joker in this deck. For systems with low numbers of units produced and high cost per unit, using powerful production computing facilities is consistent with both overall cost reduction and minimizing time to market. For high unit productions systems, however, these goals are in conflict. A hybrid solution that is possible is some cases where time to market is critically important and production runs will be very large, is to use more powerful computing resources in early production units, accepting less margin in return for rapid market entry, replacing the computer with a more highly optimized version later.

The target processor for the production prototype phase will have the same computing capability as the final production version, but substantially better diagnostic capability. It may also not fit in the same space as the final production unit.[3] Depending on the relative capabilities of the production target processor and the laboratory computer, there may be a substantial amount of software modification for this stage (or not, if the system under development is, for example, an expensive machine destined for a high production manufacturing environment). Once that has been done, the test cycle is essentially the same as for previous stages. In this case, however, the data acquisition and analysis tools will be much more primitive so more effort will have to be devoted to that process. If software changes have been made at this stage (for example, switching from floating point math to integer math) special attention will have to be paid to those parts of the control system affected by the changes.

---

[3] Note that the term "target" is used in two contexts here: for the physical system that performs the function desired by the customer, and for the computer that runs the control software, to distinguish it from the "host" computer where the software is developed.

1.13.6 Production System

Many of the problems during the development of the production version of the control system are physical. The preproduction prototyping should have validated the production software since its computing environment should be nearly identical to the production computer. The physical environment is considerably different, however. Changes in component configuration, wiring, and other details can cause unexpected problems. Diagnosis and debugging is yet more difficult than at the production prototype stage because the production computer system has been optimized for cost, so the extra diagnostic facilities have been removed.

If this design procedure has been followed rigorously, however, the commissioning process should be straightforward. Because diagnosis and debugging are so difficult at this stage, the importance of control system refinement and debugging at earlier stages is readily apparent. In addition, with concurrent design of the control and the target, many problems have been removed at the source through redesign.

1.13.7 Maintenance

Product release is the climax, but, reality is maintenance! For much commercial software, maintenance is the largest part of the total cost of the software over its commercial lifetime. The best way to control the cost of maintenance is through careful execution of the design process. Poor design and the resulting haphazard implementation inflate maintenance costs exponentially. First, the software does not work properly on release. Poor specification and testing mean that it probably only works for the few situations that were run in the development lab. Once in customers' hands, however, the software must function in circumstances that were never tested or anticipated. It is no surprise that, for many of these situations, the control system fails and the machine malfunctions. This leads to the worst kind of debugging—under the gun (and perhaps at the customer's work site).

Even under the best of circumstances there is maintenance to be done. While the number of outright failures will be much lower if the control system software has been properly designed and implemented, customers will inevitably find ways of using the product that were not part of the original plan. These will require software upgrades and bug fixes in the period immediately following product release.

The bottom line (both literally and figuratively) is that effective long-term use of control system software depends heavily on the utilization of consistent design procedures, appropriate to the level of complexity of the software. This is just as true for the mental health of the engineers responsible for the maintenance!

Chapter 2

TASKS

Tasks provide the first dimension for organization of mechatronic control software in a modular way. As noted in chapter 1, the process of defining modules is the primary engineering design activity on the way to implementation of real-time control software. The process is not unique. Not only are any number of different organizations possible for a given system, there are many different organizational schemes that can be used.

Tasks represent units of work; they are the functional subunits of the job. Tasks can often be designed by breaking a system down along both time scales and physical components. For example, consider a device with several motor-driven axes. There might be one or more tasks to handle each of the several motors; there might also be separate tasks which handle higher speed sensing operations and lower speed control operations.

In general, separate tasks are active simultaneously. They can thus be used to describe the parallelism inherent in physical systems. Internally, tasks are organized in a strictly sequential manner into *states*. States describe specific activities within each task; only one state can be active at a time. The active state controls the current activity of a given task.

The primary reason for this distinction between the nature of tasks and states is that sequential activity is a part of many mechanical systems. Even for systems that do not seem to be fundamentally sequential, such as certain types of process systems, the sequential breakdown within tasks seems to work quite well. Some systems operate on continuous media, such as paper, fibers of various sorts, and metal sheets. Other systems operate on continuous flows. The most prominent examples of continuous-flow process systems are oil refineries, but there are many other examples throughout industry. During normal operation, tasks in these types of systems tend to stay in only one or two states. However, during startup, shutdown, and exceptional conditions such as emergencies, the state of these systems changes, and the task/state structure describes their operation very well.

The sequential state structure also serves as one test for appropriate task definition. If a task is aggregated too much, it will end up with parallel operations which cannot be described effectively by states. Breaking the task into several smaller tasks solves this problem.

As the task structure is used to organize the functional definition of a problem, processes and threads organize a computational system. Processes and threads both implement parallel processing. Processes describe computational entities that do not share an address space; there can be separate processes running on one processor, processes running on independent processors in the same computer, or processes running on entirely separate computers. Threads are computational entities that share an address space but can execute asynchronously. The organization into processes and threads is done purely for performance purposes. As will be noted in a subsequent chapter, there is no theoretical necessity for such organization at all; instead, the organization serves only to allow a system to meet performance specifications when the chosen processor is not sufficiently fast to meet the specifications without using a multiprocess or multithreading structure. It does this by shifting processor resources from low priority to high priority tasks.

2.1 Example: Task Selection in a Process System

A prototypical process system is shown in figure 2-1. The sensors used in the system are not shown. The process contains three tanks that are used to store feedstock for a downstream process. It is necessary to maintain the composition of the liquid in each of the tanks, as well as the temperature. To assure an adequate supply, the level must be maintained within certain specified limits. The outflow valves are operated by the downstream process, not by the controller for this process.

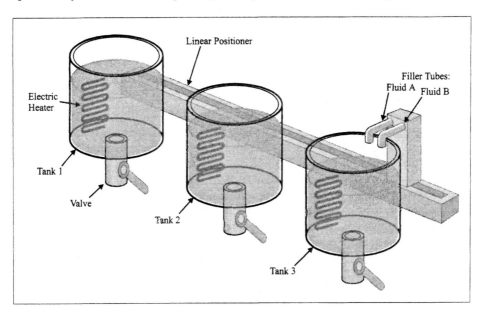

Figure 2-1. Prototype tank system

Table 2.1. Tasks used for tank system control

Name	Function
Master	controls start-up, operating mode, shutdown
OpInt	operator interface
Position	position control of carriage
Motion	gives setpoints to Position to assure smooth motion
Level	monitors and controls liquid level and mixture
Pumps	controls flowrate through pumps
Temp1	controls tank 1 temperature
Temp2	controls tank 2 temperature
Temp3	controls tank 3 temperature

Liquid is delivered to the tanks with a dual feed tube which contains separate flow controls for fluids A and B. There is only one feed tube, and it can be moved by a motor to whatever tank needs service.

The first stage in the design of the control software for this system is the selection of tasks. A possible task list is given in table 2.1. It is important to note that this particular selection is not unique.

Some of these tasks are simple and have very little internal structure. Others are more complicated. A method of notating internal task structure is necessary to provide sufficient modularity. This is done with the state specification, as will be described in the next chapter.

2.2 Tasks and the Control Hierarchy

Hierarchical organization for control systems isolates the many levels of abstraction that are needed to achieve control of a complicated engineering system. Abstractions refer to the way that a system is viewed as a function of the particular activity being examined. For example, it is not usually a good idea to consider the particular physical mechanism the pumps in the previous example use to move liquid when the concern is what volumes of liquid are needed to achieve particular concentrations of the components. Why not? Because consideration of both concerns at the same time leads to an overly complex representation which makes it difficult to arrive at a simple, efficient, and reliable implementation.

Note the weasel-word "usually." There are problems for which it seems to be necessary to consider more than one abstraction layer simultaneously. For example, if the nature of the flow transients affects the downstream process, information about those flow transients may have to be communicated to that process. Similarly, if the torque ripple in a motor (a form of vibration induced by the design of the motor) causes unacceptably large errors in a mechanical position or velocity, the controlled device may have to deal with the torque ripple in order to reduce the errors.

Crossing abstraction boundaries is expensive in both design and implementation because of the additional software complexity involved. Therefore, a redesign of the system would probably be a good idea; this might involve selecting a new motor, a different type of pump, and so on. If such a redesign turns out to be impossible for one reason or another, the problem may still be solvable but at the expense of additional time and other resources.[1]

2.2.1 Intertask Communication

The set of tasks constituting a control system must cooperate to achieve the desired system performance. That cooperation is facilitated by information flow among the tasks. We can use the details of the information flow as a major organizational aid in task design.

The general structure that we would like to achieve is one in which tasks are situated in the hierarchical structure in such a way that they communicate only (or at least primarily) with other tasks no more than one level below or one level above. A diagram can be made for a set of tasks designed in this manner showing the tasks organized by horizontal rows according to their level in the hierarchy with lines connecting the tasks that must communicate with each other. Because communication is restricted to one level up or down, the diagram is made more readable than diagrams in which communication paths can be drawn across levels.

The diagrams begin at the lowest level with the hardware to which the control system will communicate directly. This hardware usually consists of sensors and actuators. If specific hardware data processing devices are part of the system, they can be included at the next level. It is useful to include them because there is often a design tradeoff between using hardware or software for those functions (e.g., direct interpretation of quadrature signals from an optical encoder *versus* use of a hardware decoder). By including them in the task diagram, the consequences of switching back and forth between hardware and software implementation can be examined more easily.

The next higher level is the lowest level of the software. Software at this level usually has direct communication with hardware devices via the input/output (I/O) system of the computer. All tasks that have direct hardware contact should be included at this level. The nature of levels above this is application dependent. Much of the creativity in the design effort goes into figuring out how to arrange them.

2.3 Task Structure Examples

In order to illustrate the usage of the task structure more clearly, several examples of systems and their task structure are provided in this section. The examples will begin with very simple systems and progress through modestly complex ones.

---

[1] The additional time required to design and implement a system whose tasks are too complex may seem manageable, but when the time comes for *debugging* that system, the time and expense usually increase exponentially.

Section 2.3. Task Structure Examples

2.3.1 Velocity Control of a DC Motor

Problem:
A DC motor is to be operated by feedback control. Design a task structure for the motor controller.

Solution:
This is one of the simplest systems that can attract serious software design attention (the next example, temperature control, is another). It is often used as an exercise in basic feedback control classes. Figure 2-2 shows a possible task diagram for this system. A number of design assumptions are made as soon as a diagram such as this is created. It is particularly important to look for omissions. It is more difficult to add features at a later stage in the design process. It is much easier to remove features.

Probably the most important assumptions in this case concern how the system is to operate. Does the motor operate at constant speed, or variable speed? Does the motor turn on at power-up, or must it be started in some way? Is there a controlled ramp to operating speed, or does the controller just attempt to get there as quickly as possible?

Figure 2-2 shows an operator interface, which is the only possible source for much of the information about the system's mode of operation. If the operator interface is nothing but an on/off switch, all of this information must be hard coded into the control program or read from some other source such as a file or read-only memory (ROM) device. Having no direct user interface is quite common in embedded systems.

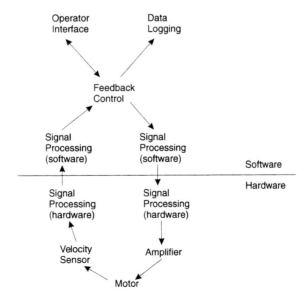

Figure 2-2. Task diagram for velocity control

On the other hand, are all of those tasks needed? If the motor uses an amplifier (motor drive) with an analog voltage input and a tachometer that produces a voltage output, there is virtually no additional signal processing required. Yet, the diagram shows both software tasks and hardware devices for that purpose. However, if the motor uses a pulse-width modulated (PWM) input and an incremental encoder for output, several of those tasks and/or devices would be needed. Some of the tasks, such as the operator interface and the data logging, might be used during development but discarded for the final production versions.

Even at this very early stage of development, a design review based on a diagram of this sort can get to the heart of many of these questions. It is not unusual to find that engineers working on one aspect of a problem have very little idea as to what kinds of expectations others in the organization (marketing, service, manufacturing ...) might have. Getting everybody on board early in the process is crucially important—many last minute (and often very expensive) efforts result from mismatched expectations.

2.3.2 Heater Control

Problem:
Design a task structure for the control system of a heater.
Solution:
Controlling a single heater is a job of similar complexity to velocity control of a DC motor. To illustrate the use of task diagrams further, a specific context will be assumed for this heater control. The heater is assumed to be part of an industrial process and represents a stand-alone piece of equipment (that is, its controller is not responsible for anything but the one heater). The entire plant is connected with a network. The command information on what the heater should be doing is sent to the heater controller via the plantwide network.

The structure for the heater control (figure 2-3) reflects this environment. In addition to the separation between hardware and software, the division between internal and external is shown. The external environment is anyplace where another computer is operating independently.

Thermal systems are usually relatively slow compared to motors, so many functions that must be done in hardware on a motor-driven system can be accomplished in software for a thermal system. Thus, it makes good sense to include the temperature signal processing and PWM as separate tasks. But why as separate tasks rather than include those functions within the feedback task? The general abstraction argument provides the answer: these tasks operate on different time scales than the feedback control task and have different functions to perform. It may be necessary to filter the temperature signal to remove noise. To do that, the temperature signal processing task could be run at a rate of 10 or 20 times faster than the feedback control task to provide the necessary filtering. The purposes of the two tasks are different as well. The act of filtering the signal to produce a better representation of the actual temperature of the control object could be necessary in temperature control systems that do not use feedback at all. For example, the

Section 2.3. Task Structure Examples 25

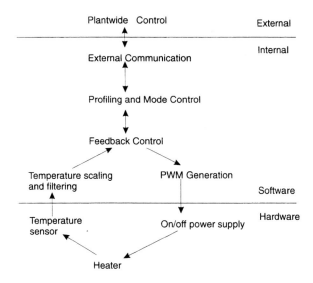

Figure 2-3. Task diagram for heater control

signal could be reported to a plantwide information system for diagnostic logging or so that it can know when the temperature has crossed a threshold such that a downstream process can be turned on.

The argument is made that for such simple systems breaking them down into this many tasks is counterproductive and a waste of time: "Simply write the program to do the job and it will be up and running before the ink is dry on these complicated diagrams." The argument is beguiling, but not true. If the tasks in this diagram are examined they will be found to operate at different time scales and have unique functions. Those functions can be performed in the absence of other tasks, as noted earlier for the filtering task. The profiling and mode control would be necessary whether or not feedback were needed to keep the temperature within tolerance. Likewise, the communication to the plantwide information system would be needed even if profiling weren't. What this implies is that by designing these tasks as modules, the design of the system can be changed without seriously affecting the other modules. As the design evolves—as other people, in other organizational functions, add their inputs to the process—changes occur. The modularity keeps the work necessary to implement those changes manageable. The documentation, the task diagram, and its accompanying explanatory material provide the basis for discussion by which the evolution of the design is guided. As will be seen in later chapters in which the means of implementing tasks in computer code is developed, keeping a task to a single hierarchical level greatly simplifies the generation of its code, making the program more reliable and maintainable. Task/state design may not be a quicker route to a program, but it is a quicker route to a reliably *working* program.

2.3.3 Toaster Oven

Problem:
Design a task structure to control the operation of a toaster oven. The oven is designed as shown in figure 2-4. There are two heating coils; the coil at the top of the oven is used for toasting and for broiling, while the coil at the bottom of the oven is used for baking. There is a single temperature sensor mounted inside the oven chamber. The operator interface consists of the following components: a display that can show time and temperature (this is an upscale appliance); a set of buttons which turn the toaster, oven, and broiler on and off; and two rotary knobs (connected to optical encoders) which are used to adjust the oven temperature setpoint and cooking time. The oven should be able to perform the following operations:

- **Toast:** Cook with both upper and lower heaters at maximum power for a specified time.

- **Broil:** Cook using the upper heating element at full power for a specified length of time.

- **Bake:** Cook using both upper and lower heaters, maintaining the temperature inside the oven at a constant value, for a given length of time.

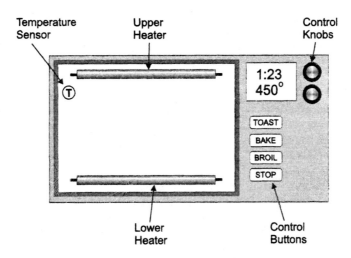

Figure 2-4. Diagram of the toaster oven

Solution:
One possible task structure for this problem is shown in figure 2-5. This application is somewhat unusual in that the user interface task may have one of the fastest timing constraints in the system. The heating elements in a toaster oven can probably

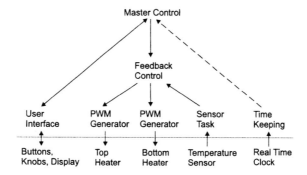

Figure 2-5. Task diagram for toaster oven controller

be adequately serviced by tasks running at a sample time on the order of a second, and the PWM task will probably need a sample time on the order of 0.1 seconds. However, the rotary dials on the user interface are connected to optical encoders, and in order to save cost these encoders will be read directly by software rather than by the special hardware used in most motion control systems. It may therefore be necessary to read the encoders at intervals of about a millisecond in order to ensure that the user's input is not missed.

Note also the dashed line connecting the sequencing task to the hardware clock. This is an indication of the special (and sometimes inconsistent) treatment given the measurement of time in embedded systems. Because time is such an important element in nearly all computing, hardware which allows real time to be measured is built into most computers. Software which allows access to measured time is usually built into the real-time programming environment. Therefore, we often take for granted that the hardware and software for measuring time will be present in our system without our having to explicitly design them in. While this is often the case, it may not be true for inexpensive embedded control computing devices. The extremely simple and inexpensive embedded control chips which are present in household appliances often do not have real-time measurement hardware built in, and so we may have to add such hardware to the system—and write and debug the software to interface with it.

2.3.4 Flexible Position Control of a DC Motor

Problem:
Design a system to control a DC motor. The configuration currently specified by the marketing department uses only a single incremental optical encoder for position and velocity feedback. But the design must allow for possible changes to other configurations such as the use of a tachometer for velocity feedback.

Solution:
Position control of motors is undergoing a change from the traditional arrangement which uses separate velocity and position sensors to configurations with only a

single position sensor, often an incremental encoder. The key to position control is stabilization. In the traditional configuration, a velocity loop provides stabilization while the position loop gets the motor to where it is supposed to be. This cascade control architecture can be implemented partly in hardware and partly in software, or entirely in software.

The single-sensor approach is taken in figure 2-6. However, the task structure will support a variety of configurations because of the separate task for handling the encoder input. That task can simply pass the current position to the feedback task, which can use some form of dynamic compensator to assure stability. It can also derive a velocity from the position which can be used in a software-based cascade configuration. If an external, hardware-based velocity loop is used, the internal control loop can be configured as a traditional position control loop with external velocity-based stabilization.

A number of explicit design decisions have been made in this case relative to the task diagram shown for velocity control. The role of the position signal processing task is now clear, based on the assumption that an encoder will be used as the feedback instrument. The decoder will normally take the high speed signal from the encoder (often up to several million transitions per second) and convert that to a position count stored in an internal register. The internal register must have finite precision, usually 8 or 16 bits. Therefore, the register will roll over (count to its maximum and then back to zero) as the motor continues to turn in one direction. The primary job for the position signal processing task is to sample that register often enough so that no roll-over information is ever lost. The information read from the position register is then converted to an internal computer representation

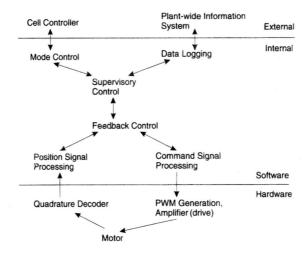

Figure 2-6. Task diagram for position control of a DC motor

with adequate precision as needed for the problem.

This task has both a very high priority and a very fast time scale—the two often go together, but not always. The high priority comes from the need for the system to maintain position integrity. That is, the internal representation of the position of the motor must always be accurate; otherwise reliable machine operation in the future cannot be guaranteed.

On the other hand, the command signal processing task in this case is mostly vestigial. Its only real function is to forward commands from the feedback control to the motor's drive amplifier. This task does serve a useful service as a placeholder in the software development cycle. When switching back and forth between simulation and physical realizations, this task takes on important simulation duties and helps to keep the feedback control task as portable as possible by insulating it from direct contact with the amplifier. In this instance, its main action is to scale the command signal from the units used by the feedback controller to the units needed by the drive.

The feedback control task is probably the easiest to understand: it implements a feedback control algorithm, getting a setpoint from the supervisory task and giving its command to the command signal processing task. This task does exactly what is described in feedback control classes.

The primary function of the supervisory control task is to provide setpoints to the feedback control task. Even in relatively simple systems, it is worthwhile separating these functions to retain the best degree of mix-and-match portability. Although it is tempting to say that the supervisory control task should be run every time the feedback control task is run, that is neither always necessary nor appropriate. For example, in an operating mode where the setpoint is constant for a long time (holding a position), there is no need to run the supervisory task as often as the feedback task. Even in a tracking situation, where the setpoint is constantly changing, the sample time requirements for the feedback control may be significantly smaller than those for the supervisor. Likely reasons for this are when stability requirements or disturbance rejection tolerances demand a faster sampling rate than is needed for smoothness in tracking the setpoint.

2.4 Simulation

Simulation having been mentioned in the previous section, a short digression on its role in control system software development is in order. Simulation in this context means a purely software realization of the entire control system along with suitable idealized models for the components falling outside the software boundary (physical elements, external computers, operator, etc.) Models of physical components become tasks in the simulated software.

It is most important to realize that the real-time, real-world control environment is a very difficult place in which to work. The various parts of the physical system "have their own minds" in the sense that they cannot be stopped and restarted at will, nor can they be frozen in time to examine a particular operating point. On the software side, all of the tasks are active simultaneously, so it is not easy to dig out cause and effect from an observed behavior. Furthermore, many of the events are happening at high speed—far higher than the person doing the debugging can keep up with.

Simulation, on the other hand, is a very user-friendly environment. All of the "can'ts" from the previous paragraph become "cans." Therefore, it is wise to get as much debugging and software verification as possible done in the friendly simulation environment rather than in the unfriendly real-world environment.

One important rule in these simulations is that the actual control system software must be used. The purpose in this case is specifically to test and debug *that* software. Thus, these simulation exercises are fundamentally different from traditional analytical simulations undertaken for the purpose of evaluating designs, optimizing parameter sets, proving operational concepts, and so on. The traditional simulations are most often performed within environments designed expressly for effective and efficient simulation, often with sophisticated numerical methods built into standard simulation software. Our control system simulations are performed in the environment which was optimized for creating control system software, not sophisticated mathematical simulations; therefore, the simulations done for control system testing will of necessity use models that are mathematically much simpler than those used in design-oriented simulations. This mathematical simplicity is acceptable because it is the operation of the control software, not the stability and performance of a control algorithm, which is being tested. It might be tempting to translate the control software to an environment which provides more powerful mathematical tools, but such translation is not a viable option because the translation process itself is time consuming and introduces new bugs. Here we can see one of the reasons that software portability is stressed so much in this text: We need to use the same software in the widely different worlds of simulation and control of physical systems.

To build a simulation, those tasks on the the task diagram which lie on the borders between hardware and software must be replaced by versions for simulation and the hardware or other components on the other side of the border must be replaced by new tasks for simulation. The computing environment is then adjusted to simulate time rather than measuring real time. The generation of "time" becomes strictly a software issue, so time granularity can be varied at will and the simulation can be frozen at any moment for full inspection of internal variables.

The simulation process is incredibly powerful. It will help to smoke out bugs and misconceptions at all levels of the control system. The productivity enhancement is substantial. Each bug, logic error, or program inconsistency that is found via simulation saves the huge time and expense of debugging in the much less friendly environment of real, physical control.

The simulation can also be extended to a real-time but simulated physical system version. While the operation is not exactly the same as it will be in the full control environment, some issues associated with timing can be studied and initial estimates can be made of computing power necessary to achieve the performance objectives.

2.5 More Task Structure Examples

The following examples illustrate the applicability of the task structure to systems of increasing complexity.

2.5.1 Coordinated, Two-Axis Motion

Problem:
Modify the task structure for the DC motor control system so that it effectively controls and coordinates two motor-driven axes.

Solution:
Closely coordinated motion of several axes is required in a large number of mechanical control problems. An example mentioned in chapter 1 is that of the masterless cam grinder. In that case, one axis controls the rotary motion of the cam being ground and the other axis controls the position of the grinder. The two must be coordinated in order to produce the correct contour on the cam. All machine tools require multiple axis coordination, ranging typically from two to six axes. Robots doing activities such as painting require multi-axis coordination in order to achieve smooth motion (pick-and-place robots and spot welding robots don't require such close coordination as do robots executing continuous-path motions). A different kind of application needing coordination happens in processes such as printing where successive stages of the press must remain synchronized so that the color-to-color registration can be maintained. The standard solution for this problem is to use a shaft connecting all of the stations for coordination (a line shaft). This line shaft is really present for information transmission purposes, but the precision with which the information is transmitted will vary with the load carried because of deflections in the shaft. Machines of this type are beginning to be built with no line shaft, using multiaxis coordination instead.

The task diagram for two-axis control looks very much like two copies of the single axis control system, with one major exception: there is only one supervisory task. This feature highlights the importance of treating the supervisory function at a higher abstraction layer than the feedback control. The supervisory control maintains coordination via the setpoints for each of the individual axes, thus clearly carrying out a function that is hierarchically above the feedback control tasks.

Feedback control is critically involved in the coordination process. For this design to succeed, the positioning errors in each of the axes must be smaller than the coordination tolerance. Setpoint profiling must be carried out in such a way that keeping the error within this limit is possible. If the controller operates in a transient mode with large error, even though the setpoints to the two axes are the same, differing dynamics or even manufacturing differences in the motors will lead to coordination errors.

2.5.2 A Washing Machine

Problem:
Design a task structure for the control of a washing machine. A simplified diagram of the machine is shown in figure 2-7.

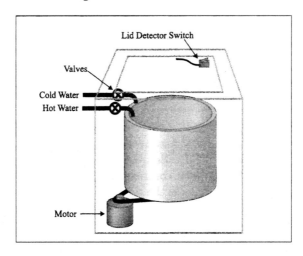

Figure 2-7. Washing machine diagram

Solution:
A possible task structure for the washing machine controller is shown in figure 2-8. In this example, there are two valve control tasks which are probably identical. This helps reduce program complexity by permitting code reuse within the program. If the development environment is an object-oriented language such as C++ or Java, the two valve control tasks would probably be implemented as two instances of the same task class. Even in a traditional procedural language, all the control functions would be the same, and the information necessary to allow each task to control a separate valve would be treated as *data*, not as *code*.

The presence of a task devoted solely to safety is significant. There are many means of making embedded control systems safer, as in any application which has

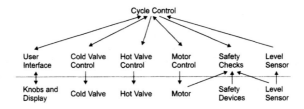

Figure 2-8. Task diagram for washing machine control

the potential of causing injury or property loss, safety must of course be a paramount concern. Hardware interlocks, such as a switch which cuts off the motor and prevents a washing machine's drum from rotating when the lid is open, should be used wherever possible. In addition, having a task which constantly monitors the system for other, less easily detected trouble can help enhance the product's safety. The washing machine safety task could detect conditions such as motor stall (by measuring the current to the motor and the speed at which the motor is operating), water overflow or leaks, and valve failure. With the addition of an inexpensive accelerometer, the safety task might even detect excessive vibration during the spin cycle to prevent the washer from dancing across the floor due to an unbalanced load of too many towels.

Chapter 3

STATE TRANSITION LOGIC

The organizational structure described in this paper is an adaptation of state transition logic[8]. This adaptation provides for implementation of the structural principles enumerated earlier. Documentation development is integrated with the software specification procedure, the software produced is inherently modular, and audit trail information can be produced automatically. By extending the transition logic paradigm to cover the entire realm of real-time requirements[7], two important results are achieved:

- The discipline associated with following a formal design paradigm is extended to the low level as well as high level software tasks.

- It becomes possible to produce portable code, that is, code which can be generated and compiled to run in a number of different real-time environments without changing the user-written portion of the code.

State transition logic is formally defined within finite automata theory[9]. As used in the design of synchronous sequential circuits, it becomes a formal, mathematical statement of system operation from which direct design of circuits can be done (Sandige[10], or any basic text on logic design). When used as a software specification tool, state transition logic takes on a more subjective cast; the transition logic specifies overall structure, but specific code must be written by the user[11],[12]. The state transition logic concept has been further specialized to mechanical system control through the specification of a functional structure for each state. This structure specifies the organization of code at the state level so that it corresponds closely with the needs of control systems.

The use of transition logic has also been based on the very successful applications of PLCs. These devices, in their simplest form, implement Boolean logic equations, which are scanned continuously. The programming is done using *ladder logic*, a form of writing Boolean equations that mimics a relay implementation of logic. In basing real-time software design on transition logic, each state takes on the role of a PLC, greatly extending the scope of problems that can be tackled with the PLC paradigm.

Depending on the nature of the problem being solved, other formulations have been proposed. For example, the language `Signal` [13] was invented for problems which have signal processing as a core activity. Benveniste and Le Guernic[14] generalize the usage to hybrid dynamical systems.

3.1 States and Transitions

State specifies the particular aspect of its activity that a task is engaged in at any moment. It is the aspect of the design formalism that expresses duration. States are strictly sequential; each task is in one state at a time. Typical activities associated with states are:

Moving: A cutting tool moving to position to start a cut, a carriage bringing a part into place, a vehicle moving at a constant velocity.

Waiting: For a switch closure, for a process variable to cross a threshold, for an operator action, for a specified time.

Processing: Thermal or chemical processes, material coating in webs.

Computing: Where to go, how long to wait, results of a complex calculation.

Measuring: The size of a part, the location of a registration mark, object features from vision input, the distance to an item.

Each state must be associated with a well-defined activity. When that activity ends, a *transition* to a new activity takes place. There can be any number of transitions to or from a state. Each transition is associated with a specific condition. For example, the condition for leaving a moving state could be that the end of the motion was reached, that a measurement indicated that further motion was not necessary, or that an exception condition such as a stall or excessively large motion error occurred.

3.2 Transition Logic Diagrams

State transition logic can be represented in diagrammatic form. Conventionally, states have been shown with circles, and transitions with curved arrows from one state to another. Each transition is labelled with the conditions that specify that transition. This format is inconvenient for computer-based graphics, so a modified form, shown in figure 3-1, is used.

This diagram shows a fragment of the transition logic for a task that controls the movement of a materials handling vehicle. The vehicle moves from one position to another, picking up parts in one position and dropping them off at another. The states are shown with rectangles; a description of the state is given inside each rectangle. The transitions are shown with arrows and the transition conditions are shown near the lines denoting transitions. The first "move-to" state shows a typical normal transition as well as an error transition, in this case based on a time-out condition.

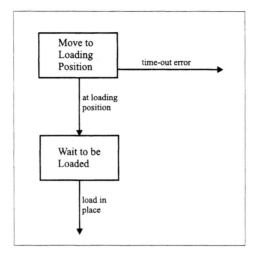

Figure 3-1. State transition diagram

Although these diagrams are not essential in using transition logic, they are an excellent visualization tool. If a task is compact enough to fit a logic diagram on a single page, the graphical description makes its function much easier to grasp.

3.3 Tabular Form for Transition Logic

The state representation of a task does not have to be graphical. Complicated tasks can have diagrams that are difficult to read and it can be difficult to automate the process of producing the diagrams from a computer.

Tabular representations can give the same information and, in some cases, are easier to deal with than diagrams. A general form for representing a state follows:

```
State Name   [descriptive comment]
- transition target #1;   reason for transition [comment]
- transition target #2;   ...
```

For the transition logic fragment shown in figure 3-1, the tabular form could be shown as:

```
Move to Loading Position [move the cart]
- Wait to be Loaded; if at loading position
- Error state; if time-out
Wait to be Loaded [hold cart in position until loaded]
- Move cart; if load in place
```

The tabular format is equivalent to the diagram and more compact. For some people it may be more readable; for others, less.

3.4 Example: Pulse-Width Modulation (PWM)

PWM is widely used as an actuation function where there is a need for a digital output to the actuating device, but continuous variation in the actuator output is desired. The need for a digital output usually arises due to concerns about cost and efficiency: Digital, or switching, amplifiers are both less expensive and more efficient than their analog, or linear counterparts. When the actuator plus target system constitute a low-pass filter, PWM can perform the function of a digital-to-analog converter by exploiting the temporal properties of the low-pass filtering. The PWM signal is usually a rectangular wave of fixed frequency with variable duty cycle (i.e., ratio of on-time to cycle time). The logic diagram in figure 3-2 shows a task to implement PWM and a typical PWM signal is shown in figure 3-3.

The PWM task has four states and will produce an effective low frequency PWM from software. The maximum frequency depends on the means of measuring time that is utilized and the timing latencies encountered in running the task. It would be suitable for actuation of a heater or perhaps a large motor, but could be too slow for a smaller sized motor whose mechanical time constant was too fast to allow effective filtering of the PWM signal.

The two main tasks, PWM_ON and PWM_OFF, turn the output on (or off) on entry and then just wait for the end of the specified time interval. COMPUTE_TIMES is

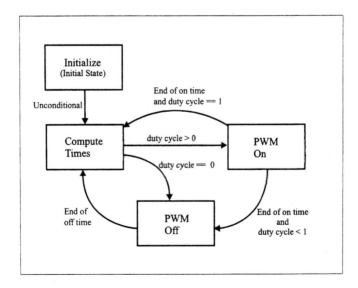

Figure 3-2. Transition logic for pulse width modulation

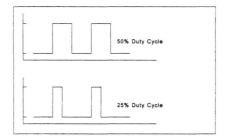

Figure 3-3. Pulse width modulation: two duty cycles

active once per cycle to find the appropriate on and off times in the event that the duty cycle has been changed. The transition conditions take account of the two special cases—duty cycles of 0 (never on) and 1 (always on) in addition to the usual on-off switching operation.

This example shows the use of transition logic for a task that is quite low level computationally. Unlike a conventional implementation of such a task, the details are readily apparent from the transition logic and reference to the code (not shown here) need only be made to confirm that it is an accurate implementation of the transition logic.

3.5 Transition Logic for the Process Control Example

The diagram of the process system of the previous chapter is repeated in figure 3-4. The control objective is to maintain the mixture, temperature, and level in each of the tanks.

Of the tasks defined for this system, the level task is probably the most complex. Its job is to figure out which tank needs attention at any time, ask for carriage motion, and initiate the filling action. The transition logic for this task is shown in figure 3-5.

This task functions largely in a supervisory role. It examines the current status of each of the tanks, and then sends out commands to provide the service it deems necessary. Other tasks are responsible for actually carrying out those commands. This state diagram is typical of the level of detail that is used. The Monitor state is identified as determining which tank needs service. How this is determined is not specified in the state diagram. That is a well-defined function, and should be specified in the auxiliary documentation accompanying the state diagram (or tabular state representation). When code for this function is written, there will be a very clear path indicating when and how the code is used, and an explanation of what the code should be doing. Other states are simpler and probably do not need much if any auxiliary documentation.

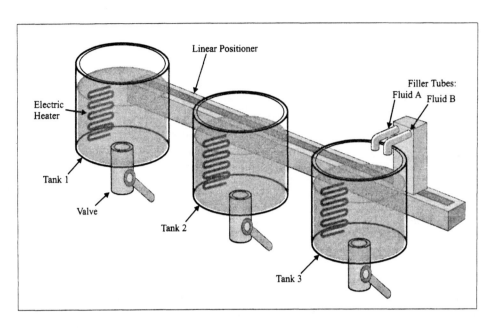

Figure 3-4. Prototype process system

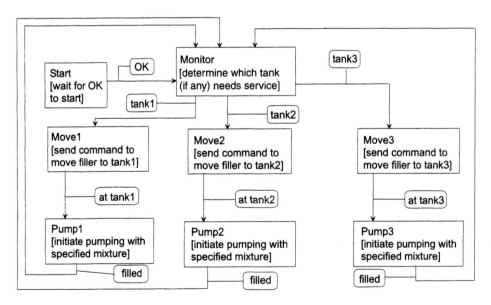

Figure 3-5. Transition logic for the level task

3.6 Nonblocking State Code

A major feature of PLCs contributing to their success as industrial control components has been that the logic function is continually scanned. PLCs are programmed in a form of logic notation called ladder diagrams. In programming ladder logic, the programmer does not deal with program flow control, as must be done when using conventional programming languages. As long as the ladder is active, it is scanned repeatedly, so the user only has to be concerned with the fundamental performance issue of whether the scan rate is adequate for the particular control application.

Transition logic design is based on this same scanning principle for execution of state-related code. In order to achieve greater functional flexibility than is possible with ladder diagrams, however, standard sequential languages are used for coding. To implement a scanning structure with algorithmic languages requires the restriction that only *non-blocking* code can be used. Non-blocking code is a section of program that has predictable execution time.

The execution time for *blocking* code cannot be predicted. This definition does not say anything about how long that execution time is. For now, it is enough that it be predictable. How long the execution time is for any specific piece of code is a performance issue which will be dealt with later.

Examples of blocking code in the C language include, for example, the `scanf()` function call used to get keyboard input from the user. This function only returns when the requested input values have been typed; if the user goes out for coffee, the function simply waits. Likewise, the commonly used construction to wait for an external event such as a switch closure,

```
while (inbit (bitnum) == 0);
```

is also blocking. If the event never happens, the `while` loop remains hung.

This restriction to nonblocking code does not cause any loss of generality at all. Quite the contrary, the transition logic structure is capable of encoding any kind of desired waiting situations shown in the examples. By encoding the wait at the transition logic level rather than at the code level, system operations are documented in a medium that an engineer or technician involved in the project can understand without having to understand the intricacies of the program.

3.7 State-Related Code

The transition logic metaphor encourages the use of modular software by associating most of the user-written code with states. Although this level of organization is adequate for simple projects, an additional level of organization is necessary in most cases. To this end, an additional formal structure of functions is established for the state-related code. Two goals of modular code writing are thus fulfilled:

- Sections of code are directly connected to identifiable mechanical system operations.

- Individual functions are kept short and easily understood.

These functions are defined around the operational assumption of a scanned environment—the code associated with a state is scanned, that is, executed on a repeated basis, as long as that state is the current state for a task. Code associated with noncurrent (inactive) states is not executed at all.

For each state, the following functions are defined:

- **Entry:** Executed once on entry to the state.

- **Action:** Executed on every scan of the state.

- **Transition test:** Executed after the action function on every scan of the state.

In some cases there is code that is unique to a specific transition. This is subtly different from code that is associated with entry to a state, because the entry to the state could be via one of several different transitions. That code can be placed in the transition test function and only executed when a transition is activated, or it can be in a separate exit function that is associated with a specific transition.

This structure enforces programming discipline down to the lowest programming level. All of these functions must be nonblocking, so test functions, for example, never wait for transition conditions. They make the test, then program execution continues.

Relating code to design-level documentation is also enforced with this structure. Transition logic documentation for each task identifies states in terms of what the mechanical system is doing. Not only is the code relevant to that state immediately identified, the code is further broken into its constituent parts.

3.8 State Scanning: The Execution Cycle

The state scan is shown in figure 3-6. In addition to establishing the execution order for the state-related functions, it also provides the basis for parallel operation of tasks.

Each pass through the cycle executes one scan for one task. If this is the first time the scan has been executed for the current state, the entry function is executed. The action function is always executed. Then the transition test function is executed. If it determines that a transition should be taken, the associated exit function or transition-specific code is executed and a new state is established for the next scan.

Behind the execution details, there must be a database of task information. Each task must have data specifying its structural information—that is, all of the states and transitions, task parameters such as priority, sample time, and so on, and transient information such as present state and status.

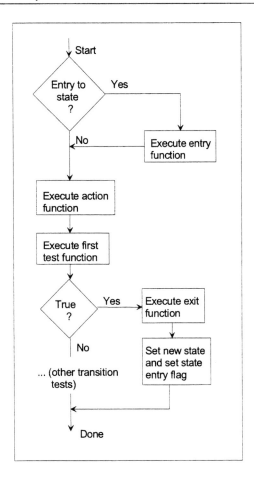

Figure 3-6. State Scan Cycle

3.9 Task Concurrency: Universal Real-Time Solution

Tasks, as noted earlier, must operate concurrently. This structure provides for parallel operation of tasks even in the absence of any specific multitasking operating system or dispatcher. Because all of the state functions are nonblocking, the scan cycle itself is also. It can therefore be used to scan each active task in succession. After finishing with all of the tasks, the first task is scanned again. This guarantees fully parallel operation of all tasks.

This method of dispatching, called cooperative multitasking, will be an adequate solution if the total scan time for all tasks is short enough to meet the system timing constraints.[1] If cooperative multitasking is not adequate, a faster computer must be used, or other dispatching solutions must be found. These will be discussed later.

The methodology discussed thus far therefore presents a universal real-time solution. It is capable of solving all real-time problems, without any special real-time constructs or operating systems, if a fast enough computer is available. All of the usual real-time facilities such as semaphores, task synchronization, and event-based scheduling can be implemented using the formalism of transition logic, with all code nonblocking.

These various facilities of real-time operating systems serve two main functions: making sure that high priority activities are serviced before low priority activities, and providing protections for problems that arise because of the asynchronous nature of real-time control software. A common problem in control system software is that while there might be enough programming power overall, high priority activities do not get enough attention. This is an obvious potential problem with the cooperative multitasking described earlier because all tasks are treated equally. One solution to this dilemma is to use a faster computer. However, if the use of a fast enough computer is not practical, use of preemptive dispatching based on interrupts can be implemented. Interrupts do not make the computer any faster; in fact, interrupts have overhead that actually reduces total computing capability. What interrupts do accomplish, though, is to allow higher priority activities to preempt lower priority activities so as to avoid the use of a faster (and more expensive) computer when the real problem is honoring priority rather than total computing capability. To do this, the transition logic paradigm must be extended to include the designation of task types and library code must be included to allow integration of transition logic based code with a variety of real time and/or multiprocessor environments. This extension will be discussed in chapter 8.

---

[1] The term "scheduling" is often used synonymously with dispatching; in this material, "scheduling" will usually be used for a planning activity concerned with meeting execution time targets, while "dispatching" will be used for the part of the program that actively directs task execution.

Chapter 4

DIRECT REALIZATION OF SYSTEM CONTROL SOFTWARE

The task and state specification for a project, along with the auxiliary documentation, should constitute a complete engineering specification for the operation of a mechatronic control system. The realization stage of the project is the translation of the specification into computer code. The computer code is ideally constructed independently of specific computational environments. It is ideal in the sense that choice of language and level of code formality will have an effect on computational environments into which the code could be ported. For example, if the language C++ is chosen to prototype a problem that will ultimately be implemented in a minimally equipped microcontroller, the final production code may well not be in C++. Even in these situations, however, because the primary technical description of what the code does is in the task/state specification, code translation from one realization to another should be relatively simple.

This chapter will deal with realization of the code using the simplest level of formality and implementation of that code in the simplest possible computational environments—those that provide at most the ability to tell time but do not support any form of preemptive execution. It is the first step in specification of a method of dispatching based on tasks which are grouped according to priority. The only facilities expected are a compiler for the target language and an operating environment to run the program.

4.1 Language

Computer languages are the source of fierce debate among their devotees. While many of the debated points have considerable merit, the approach taken here is that the important information lies in the level of abstraction *above* the code, so the particular choice of language is mostly a matter of convenience and access.

A wide variety of implementations are possible once the task/state description is done. Depending on the complexity of the problem and the stage in the design process, the implementation might be done in a relatively informal manner, or it could be accomplished in a formally defined environment. The software tools available will also vary widely. For that reason, sample solutions are presented in several languages.

The simplest implementations environment from the language point of view use Matlab. Matlab is simple from the language structure perspective but provides a very rich environment for simulation and analysis. This is exactly what is needed for early design and feasibility stages. Matlab is highly interactive and is a good environment for debugging, but it has very slow execution speed for the types of problems that are common in mechatronics. However, even at the slow execution speed, with a fast computer it could be used for the actual realization of process control systems.

The other languages discussed, C++ and Java, are both derivatives of C. C is by far the most commonly used high-level language for control of mechatronic systems. C++ has additional structure beyond that of C and that structure is extremely useful in imposing a programming style that is much more effective in capturing the task/state structure used for the engineering. Java is a language similar to C++. It simplifies some of the more error-prone features of C++ (pointers and memory leaks, for example) and adds syntax for many computational elements that are not part of C++ syntax (for example, networking and GUIs). C++ and Matlab are discussed in this chapter; Java is discussed in chapter 13.

Because of its very strong internal structure, C++ will be utilized to build environments in which task and state formalisms are already defined so the user has only to fill in the application specific code. For environments which only support C, this structure can be reproduced in C in either a formal or *ad hoc* manner.

The highly formal C++ solutions have the greatest degree of portability since the underlying software that connects the user code to the computational environment can be treated as a utility and need only be written once for each environment. The application code does not need to change at all. On the other hand, a full C++ compiler is necessary in each target environment. The less formal solutions, usually implemented in C, require more change for porting, but only require a C compiler or, sometimes, no more than a subset-C compiler.

Implementation in almost any other language is possible. It should be easy to translate the examples in Matlab, C, or C++ into almost any language, including assembly language (if one has sufficient patience).

Code for short programs or fragments of longer programs will be shown in the text. Full source code is available from the archive collection associated with this text.

4.2 Time

Almost all mechatronic control programs need access to explicit values of time. In the direct implementations described in this chapter all of the decisions tasks need to make about time are done explicitly, within the task. These calculations definitely need to be able to know the current time. Within the control elements of the program as well, time is often explicitly used—for example, in scheduling or dispatching a certain amount of time to a particular operation. Simulation programs only need their internal, simulated time. There is no real time.

The simplest way of generating real time for actual realizations of control programs is through calibration. This requires no specific computer hardware nor software utilities. It is accomplished by establishing an internal program variable for time, and incrementing that variable at a fixed spot in the program (much as is done in a simulation program). Calibration is done by running the program for a long enough time to exercise as much as possible of its functionality and for a long enough time to get a reliable measure of actual run time. The value by which time is incremented each pass is then adjusted so that the program's internal time matches real time.

This method is not very accurate, but it is extremely simple to implement. It is quite good when what is needed is a rough knowledge of time but high accuracy is not required for either very short or very long time periods. The calibration must be redone whenever the program is run on a different processor, whenever the overall load on the processor is changed by some external circumstances, or whenever the program is changed. The calibration is very easy to do, however.

To go beyond the accuracy available with the calibrated method, an actual physical clock must be used. This normally consists of a highly stable crystal oscillator whose output drives a counter. Details of how the clock is implemented will be presented in a later chapter. For the purposes of this chapter, it is assumed that the current time is always available via a function call. That time value will normally be a measure of the time since the program was started, but sometimes the scale will have an arbitrary origin.

4.3 Program Format

As a means of organizing control programs, the suggested format is that each task be represented by a separate function and, where possible, each task's function is in a separate file. This assures maximum portability and makes it easy to match code to documentation. The code which implements states—action, entry, test, and exit functions—can be defined in line or defined in separate functions, depending on the complexity of the state logic.

In addition to the task and state related code, there is usually some form of "main" function that is used to set the program up, initialize variables, and then start the control operation. The form taken by this part of the code depends strongly on the development environment.

4.4 Simulation

Simulation provides an extremely valuable environment for the initial testing and debugging of control programs. The most important aspect of using a simulation is that the basic software can be tested in an environment that is convenient and efficient to use and in which it is relatively easy to find bugs. The simulation environment is a conventional, strictly synchronous computation environment so errors can normally be reproduced by running the program again. This is often not the case when the control is operated in its target environment because the control program and the target physical system operate asynchronously.

It is very tempting to skip the simulation step in system development and go straight to lab prototype. The temptation arises because it takes significant resources to construct the simulation. Would these resources be better used to build the lab prototype? In general, *no*. Debugging in the simulation environment is so much more efficient than in the prototype environment that the investment in a simulation pays off rapidly.

4.5 Simulation in Matlab

All of the files referred to in this section are found in the archive file grp_mtlb.zip. The grp refers to these program being part of the group priority method of program construction. The zip file contains directories for each of the examples.

4.5.1 Templates for Simulation Using Matlab

This template gives a model for direct implementation of the task/state model of control systems.[1] "Direct" indicates that no software base is used in building the control program other than the language facilities themselves. The model is thus applicable to virtually any computing environment and any computing language. Matlab is a convenient environment for simulation, and the language is simple enough so that the results can easily be translated to any other desired target language. This section will thus serve as a template for applications in any language other than C, C++, or Java. As computer speeds increase, Matlab itself could become a convenient real-time environment for prototyping; otherwise a translation must be done before control of the actual target system can be accomplished.[2]

The template in the example code is in two parts. The first part is the initialization and runtime control portion of the program. This is done here as a Matlab script file, although it could be implemented as a function as well. The primary control loop has three major parts: the output section, the execution of the tasks, and the simulation of the continuous-time part of the problem.

The other part of the template is the task file. It shows the structure of a typical task. Each task would have a file matching the template in form.

---

[1] archive: grp_mtlb.zip, directory: template
[2] Matlab now supports real-time usage in certain circumstances: see http://www.mathworks.com for more information

Section 4.5. Simulation in Matlab

Note that in some cases the word processor will wrap long lines in the listing to a second line. This could result in line fragments that are not syntactically correct—for example, comments that have been wrapped put the last part of the comment on a line with no comment indicator (% in Matlab).

Initialization and Runtime Control

```
% Template for Matlab mechatronics control programs
% This template is for a single-thread, sequential scheduler
% File:tpl_seq.m
% Created 8/1/95 by DM Auslander


glbl_var     % Read in global definitions -- because each function
             %in Matlab requires a new file, this is a more compact
             %way to share variables

% System setup and initialization
% Put initial values for all variables here

% Simulation parameter values
tfinal = 30;
del_t = 0.2;         % Simulates the minimal increment of computing time
ept = del_t / 100;   % Used to check for round-off errors

del_tsim = 0.6;      % Simulation discretization interval. Must be >= del_t

% Use this to control the number of output values
t_outint = 0.5; % Time interval for outputs. Usually greater than del_t
% This is particularly important for the student edition!
nouts = ceil(tfinal / t_outint);
yout = zeros(nouts + 1,4);   % Set this up for proper output dimension
tout = zeros(nouts + 1,1);
t_outnext = 0;               % Time at which next output will be copied
iout = 1;                    % Index for outputs
tstep = 0;                   % Running time

% Set up the options for the differential equation solver
% (if one is being used).
% Several simulation options are possible. Odeeul is a function
% that does a one-step Euler integration. It is very efficient, but very
% dangerous mathematically because it has absolutely no error control
% Another good option is to use the functions in Odesuite, a new
% Matlab integration package. The version used here has been modified
% with a parameter 'hfirst' so it can be used more efficiently in this
% hybrid (discrete-time/continuous-time) environment.

ode_opts = odeset('refine',0,'rtol',1.e-3); % Differential solver options
% 'refine' = 0 causes the ode solver to output only at the end

% Make up a task list using the string-matrix function, str2mat()
% Note that this can only take 11 strings in each call, so it must be called
% multiple times for more than 11 tasks
% Each name represents a function (file) for that task
```

```
t1 = str2mat('task1','task2','task3');
tasks = str2mat(t1);
nn = size(tasks);
ntasks = nn(1);        % Number of rows in 'tasks'

% Set initial states
Stask1 = 0;            % A state of 0 is used as a a flag to the
Stask2 = 0;            % state to do whatever initialization it needs
Stask3 = 0;
% Also set up the state vector for ode simulation here if one is being used

% Initialize all the tasks
for j = 1:ntasks
    tsk = tasks(j,:);  % name of j-th task
    feval(tsk);        % No arguments - task takes care of state
                       % and when it runs internally.
end

i_audit = 1;           % Index for transition audit trail
tbegin = get_time;     % Time at the beginning of simulation step
                       % Step out the solution
while tstep <= tfinal
    % Copy output values to the output array
    if (tstep + ept) >= t_outnext
        t_outnext = t_outnext + t_outint;   % Time for next output
        yout(iout,1) = out1;
        yout(iout,2) = out2;
        yout(iout,3) = out3;
        yout(iout,4) = out4;
        tout(iout) = tstep;                 % Record current time
        iout = iout + 1;
    end

    % Run the control tasks

    for j = 1:ntasks
        tsk = tasks(j,:);       % name of j-th task
        feval(tsk);             % No arguments - task takes care of state
                                % and when it runs internally.
        inc_time;               % Increment time
        del = get_time - tbegin;  % Time since last simulation
        if del >= del_tsim        % If true, do a simulation step
            tbegin = get_time;
            [tt,xx] = odeeul('proc_sys',del,x,ode_opts);
            x = xx;              % Update the state vector
        end
    end
    tstep = get_time;    % Update time
end
% Record the final output values
yout(iout,1) = out1;
yout(iout,2) = out2;
yout(iout,3) = out3;
yout(iout,4) = out4;
tout(iout) = tstep;    % Last recorded time
```

Section 4.5. Simulation in Matlab

The most important part of this template is the loop that executes the tasks,

```
for j = 1:ntasks
    tsk = tasks(j,:);           % name of j-th task
    feval(tsk);                 % No arguments - task takes care of state
                                % and when it runs internally.
    inc_time;                   % Increment time
    del = get_time - tbegin;    % Time since last simulation
    if del >= del_tsim          % If true, do a simulation step
        tbegin = get_time;
        [tt,xx] = odeeul('proc_sys',del,x,ode_opts);
        x = xx;                 % Update the state vector
    end
end
```

For each pass through the major loop, this inside loop will cause all of the tasks to be executed once, thus implementing the basic multitasking. This is considered the basic unit of computation, so the simulated time is incremented for every scan. Execution of the simulation is also done inside this loop. To improve efficiency a bit, the simulation can be executed on a longer time scale than once every minimal time step. The ODE-solver shown here is an extremely simple one-step Euler integrator. It was written specifically for these applications, since the basic time step unit for control is often small enough for the Euler integrator to be effective. It is not a general-purpose integration function. Intergration routines based on a modified version of the Matlab ode-suite are used in other examples.

For each task, this scan will cause execution of the state code associated with the current state for that task, as shown next.

Template for a Task

```
function task1
% Template for a typical task function (which has to be
% a separate file with the same name in Matlab)

glbl_var % Global definitions

% The 'next-state'is maintained as a separate variable so that the
% initialization can be handled correctly and so a test can be
% constructed to tell whether the entry function of a state
% should be run. The next-state variable must be static so it
% is remembered across function invocations. The 'global' designation
% is the only way Matlab has of making a variable static.
% The default value of the next-state variable is -1, indicating
% that no state change has taken place.

    global S_task1_next  % Next state for this task (other local statics
                         % can be listed here).
                         % These aren't really global, but do have
                         % to be 'static'; Matlab doesn't have a 'static'
                         % declaration so this will have to do!
```

```
if S_task1 == 0
    % Initialization section - occurs once only
    S_task1 = 1;        % Make sure this section is not executed again!
    S_task1_next = 1;   % First state.
                %The default is -1 which indicates stay in same state
    return;
end

% This task runs at specific times. If current time has not reached
% the scheduled run time, just return.
if tstep < Ttask1
    return;
end

Ttask1 = Ttask1 + Tdelta;   % Time for next run -- in this case
                % the task runs at regular intervals, but that needn't
                % be true all the time.

if S_task1_next ~= -1
    % There has been a state change (including self-transition)
    [i_audit,trans_trail] = audit(2,S_task1,S_task1_next,...
    tstep,trans_trail,i_audit);
    % Record this transition
    S_task1 = S_task1_next;
    run_entry = 1;
    S_task1_next = -1; % Default - stay in same state
else
    run_entry = 0;
end

% Run the code associated with the current state
if S_task1 == 1
    % Entry section
    if run_entry    % Only run this on entry to the state
                    % code for entry section goes here (if any)
    end
    % Action section
    % code for action section goes here (if any)
    %Test/exit section
    if s1_test      % Test for first possible transition
                    % this could be a separate function or
                    % in-line code.
    % A new state has been requested
    S_task1_next = 2;   % Specify the next state

    % Exit section
    % code for exit section (if any)
end

% Similar code goes here for additional transitions
% associated with this state.

% Code for another state
elseif S_task1 == 2
    % ... Repeat this pattern for as many states as there are
    % in this task
```

Section 4.5. Simulation in Matlab

```
else    % Check for a state variable value that doesn't
        % correspond to a defined state.
    error('Task: Task1 -- unknown state encountered');
end
```

Every time a transition between states is detected, a call is made to the audit function. This keeps track of all the transitions. The audit function is:

```
function [i_audit,trans_trail] = audit (task, from_state, to_state, ...
                                        time, trans_trail, i_audit)
% Record state transitions

i = i_audit;
trans_trail(i,1) = time;
trans_trail(i,2) = task;
trans_trail(i,3) = from_state;
trans_trail(i,4) = to_state;
i_audit = i_audit + 1;
```

4.5.2 Simulation of PWM Generator

The Matlab templates in the examples are used to build a simulation of the PWM generator introduced earlier.[3] The simulation includes two tasks: one that produces a command signal specifying the desired duty cycle and a second task that generates the PWM itself. The first task is very simple, and has only a single state. The second, the PWM generator, has the state structure in figure 4-1 (repeated from previous chapter). Excerpts of the code for this sample follow.

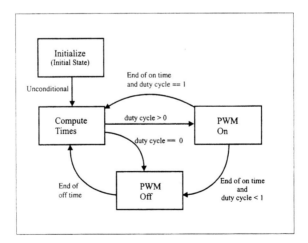

Figure 4-1. Transition logic for pulse width modulation

---

[3]archive: grp_mtlb.zip, directory: pwm_samp

The main script file follows the template model very closely and will not be displayed here. This simulation generates the PWM signal, but has no target system. It thus does not utilize the ODE functions.

The command task has only one state. That state computes a sinusoidal duty cycle command signal. The code for its sole state is:

```
% This task has only one state -- it produces a command duty cycle as
% a function of time.

% Run the code associated with the current state
if Spwm_cmd == 1
    % Entry section
    % No entry section for pwm_cmd

    % Action section
    duty_cycle = 0.5 * sin(tstep * 2 * pi / cmd_period) + 0.5;
    % Result is in range 0->1
    %Test/exit section
% This task has only one state - no transitions!
```

The PWM generator task follows the state diagram very closely. The only minor difference is that the INITIALIZE_PWM state is absorbed into the initialization section. The code for this follows:

```
if Spwm_gen == 1 % **Compute_Times**
    % Entry section
    if run_entry     % Only run this on entry to the state
                     % code for entry section goes here (if any)
        duty_cycle = limitval(duty_cycle,0,1);
        end_on_time = duty_cycle * pwm_period + tstep;
        end_off_time = pwm_period + tstep;
    end
    % Action section
    % code for action section goes here (if any)
    % No action function for this task
    %Test/exit section
    if duty_cycle > 0
        Spwm_gen_next = 2;  % Transition to PWM_On

    % Exit section
    % code for exit section (if any)
    % No exit section for this transition
    elseif duty_cycle <= 0
        Spwm_gen_next = 3;  % Transition to PWM_Off

    % No exit section for this transition
    end

elseif Spwm_gen == 2    % **PWM_On**
    % Entry section
    if run_entry    % Only run this on entry to the state
    % code for entry section goes here (if any)
        pwm_out = 1;    % Turn on the PWM output
    end
```

Section 4.5. Simulation in Matlab

```
    % Action section
    % code for action section goes here (if any)
    % No action function for this task
    %Test/exit section
    if (tstep >= end_on_time) & (duty_cycle < 1)
        Spwm_gen_next = 3;  % Transition to PWM_Off
    % No exit section for this transition
    elseif (tstep >= end_on_time) & (duty_cycle >= 1)
        Spwm_gen_next = 1;  % Transition to Compute_Times
    % No exit section for this transition
    end

elseif Spwm_gen == 3 % **PWM_Off**
    % Entry section
    if run_entry       % Only run this on entry to the state
        % code for entry section goes here (if any)
        pwm_out = 0;       % Turn on the PWM output
    end
    % Action section
    % code for action section goes here (if any)
    % No action function for this task
    %Test/exit section
    if tstep >= end_off_time
        Spwm_gen_next = 1;  % Transition to Compute_Times
        % No exit section for this transition
    end

else       % Check for a state variable value that doesn't
% correspond to a defined state.
    fprintf('Uknown state: %f\n',Spwm_gen);
    error('Task: pwm_gen -- unknown state encountered');
end
```

Running this simulation produces the results shown in figure 4-2. The top graph shows the duty cycle command, which is the product of the command task. The bottom graph shows the PWM signal.

As the duty cycle rises, the proportion of the time that the output spends at its high value increases. The frequency of the signal remains constant.

One of the major validation and debugging tools associated with the use of state logic is the transition audit trail that is produced. As noted, the array constituting the audit trail comes from the calls to the audit function embedded in the task functions. For example, the function call from the PWM generation task looks like:

```
[i_audit,trans_trail] = audit(2,Spwm_gen,Spwm_gen_next,...
                     tstep,trans_trail,i_audit);
```

The first argument is the task number, so the task associated with the transition can be identified. The next two give the from state and the to state, followed by the time of the transition, the array to store the audit information, and then the array index. The audit trail from the PWM simulation follows. The first two transitions correspond to the initialization process (at time = 0). Following that, it can be seen that all of the transitions refer to task #2, the PWM generation task. This is because task #1, the command task, has only one state and thus no further transitions.

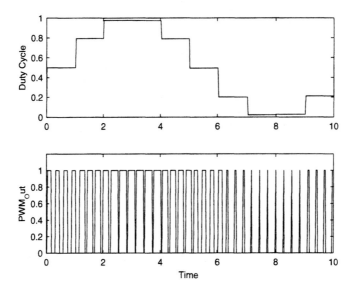

Figure 4-2. PWM test results

The entry at time = 0.12 is the transition from state 1 (Compute_times) to state 2 (PWM_On). This is followed by a transition to state 3 (PWM_Off) and 0.62 and then a transition back to state 1 (Compute_times) at time = 1.12. Since the time difference between these transitions is each 0.5, the output at that point is a 50 percent duty cycle, which can be confirmed from the graph.

Although the numbers-only format is not the easiest to read, Matlab does not have very sophisticated text-handling functions so this is a reasonable compromise.

```
trans_trail =
(Time      Task    From-state To-state)
      0    1.0000   1.0000    1.0000
      0    2.0000   1.0000    1.0000
 0.1200    2.0000   1.0000    2.0000
 0.6200    2.0000   2.0000    3.0000
 1.1200    2.0000   3.0000    1.0000
 1.2200    2.0000   1.0000    2.0000
 2.0200    2.0000   2.0000    3.0000
 2.3200    2.0000   3.0000    1.0000
 2.4200    2.0000   1.0000    2.0000
 3.4200    2.0000   2.0000    3.0000
 3.5200    2.0000   3.0000    1.0000

 3.6200    2.0000   1.0000    2.0000
 4.6200    2.0000   2.0000    3.0000
 4.7200    2.0000   3.0000    1.0000
 4.8200    2.0000   1.0000    2.0000
 5.6200    2.0000   2.0000    3.0000
```

Section 4.5. Simulation in Matlab

```
5.9200    2.0000    3.0000    1.0000
6.0200    2.0000    1.0000    2.0000
6.5200    2.0000    2.0000    3.0000
7.1200    2.0000    3.0000    1.0000
7.2200    2.0000    1.0000    2.0000
7.3200    2.0000    2.0000    3.0000
8.3000    2.0000    3.0000    1.0000
8.4000    2.0000    1.0000    2.0000
8.5000    2.0000    2.0000    3.0000
9.4000    2.0000    3.0000    1.0000
9.5000    2.0000    1.0000    2.0000
9.8000    2.0000    2.0000    3.0000
```

4.5.3 Simulation of Three-Tank Process System

The three-tank process system is more complicated than the PWM system since it encompasses more layers in the typical control system hierarchy.[4] It also has some interesting decision-making since the filler nozzles have to be moved from tank-to-tank. The code is too long to be repeated here in its entirety, so excerpts will be given.

The diagram of the system is repeated, showing the tanks and the filler carriage (figure 4-3).

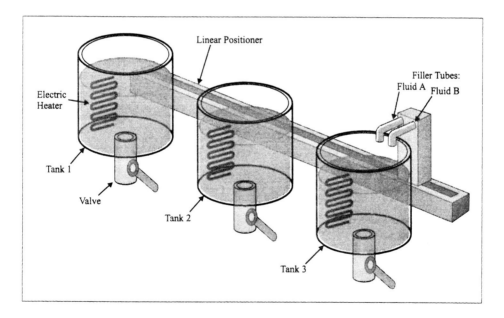

Figure 4-3. Prototype process system

---

[4]archive: grp_mtlb.zip, directory: proc_sys

The Matlab simulation uses three tasks (the heater controls are not yet simulated):

- Carriage position feedback control
- Carriage motion supervisory control
- Tank level and concentration control

The Matlab simulation also does not have any operator interface. The lack of an interactive interface is a common feature of simulations.

The state diagram of the level control task was already given, and is repeated (figure 4-4).

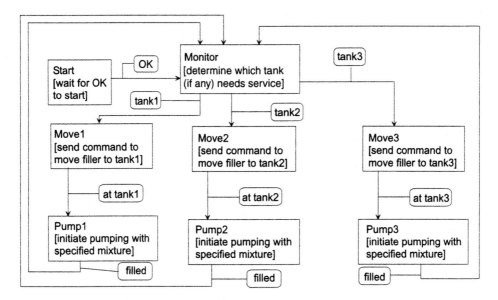

Figure 4-4. Transition logic for the level task

The Monitor state is the most interesting one in the Level task. It decides how to make use of the filler carriage, which is a singular resource for the system. It then sends commands to the tasks that perform the actual actions. As such, it is a form of supervisory task also since it determines setpoints and commands for other tasks. The code that implements the Monitor task is:

```
if Slevel == 1 % Monitor
    % Figure out which tank needs service
    % Needing service is determined by the tank with the lowest level
    % Service is needed if the lowest level is less than (high + low)/2
```

Section 4.5. Simulation in Matlab

```
    % No entry section
    % No action section
    % Test/exit section
    imin = 1; hmin = h1;
    if h2 < hmin, imin = 2; hmin = h2; end
    if h3 < hmin, imin = 3; hmin = h3; end
    if hmin < (hhigh + hlow) / 2
    % A tank needs filling
      if imin == 1, slevel_next = 2;    % to Move1
      elseif imin == 2, slevel_next = 4; % to Move2
      elseif imin == 3, slevel_next = 6; % to Move3
      end
end
```

A very simple decision algorithm is used here: if any tanks are below their specified mean heights (in this case, the same for all tanks), those tanks are serviced in numerical order. Furthermore, no other tanks are serviced until the tank currently being serviced has reached its specified height.

The carriage-motion task is the other supervisory task. Its purpose is to generate setpoints to the position control task so that the motion of the carriage will be smooth. In this case, this is done by generating a trapezoidal motion profile – constant acceleration, followed by constant velocity, followed by constant deceleration. The state logic for this is shown in tabular form:

```
Wait for Next Move (do nothing until a command is received)\\
- Accelerate if motion command received\\
Accelerate (Constant accel to cruise velocity)\\
- Decelerate if stop distance reached\\
    (a short move will not have any cruise section)\\
- Cruise if cruise velocity is reached\\
Cruise (Maintain constant velocity)\\
- Decelerate if stop distance reached\\
Decelerate (Constant decel to stop)\\
- Wait for next move if velocity reaches zero\\
```

A typical move contour is shown in figure 4-5. The velocity has sharp corners when moving in and out of Cruise, but the position curve is smooth. This profile is satisfactory for many move profiles, but still smoother ("jerkless") profiles can be generated if necessary.

A typical simulated system performance is shown in figure 4-6. The top set of lines is the height of liquid in each of the three tanks. The bottom solid line is the position of the carriage. As long as the height in all three tanks remains suitably high, the carriage remains at its initial position ($x = 0$). When the tank shown by the dashed line reaches a critical height, the carriage starts moving. The filling is delayed by the time the carriage is in motion. Filling starts as soon as the motion stops, as can be seen from the upward slope in the liquid height curve.

The effect of the motion supervisory task is seen in the smooth transition of the carriage from moving to stopped.

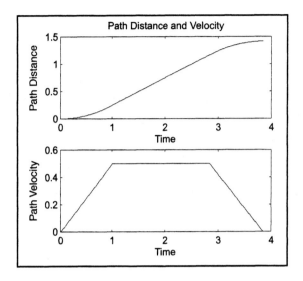

Figure 4-5. Trapezoidal velocity profile

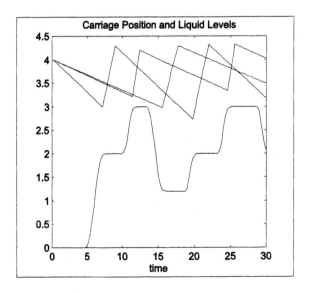

Figure 4-6. Process system behavior

4.6 Intertask Communication

Communication of information between tasks is accomplished in the Matlab environment through the use of the `global` definition. While this is adequate for Matlab—and convenient because the Matlab environment becomes awkward if large numbers of functions are defined—it does not provide a general enough mechanism for intertask communication. The reasons for this involve both data integrity and program portability, and will be discussed in more detail in a later chapter.

The mechanism for writing programs in the direct and simply structured manner used in this chapter but still preserving easy portability to more complex environments requires two steps:

- Associating variables with tasks rather than with the whole program.

- Explicitly identifying all information transfers between tasks.

The simplest way of implementing this is: for every variable in a task that is needed in other tasks, define an associated variable that is used only for transferring information. Using variable names that begin with `x_` (meaning exchange) will help identify them quickly when looking at code. Then, for each such variable define a `get_xxx()` function that other tasks can use to get the value of that variable when it is needed. For efficiency, variables that are always needed as a group can share a single function.

The exchange variables are set to the value of their corresponding local variables at a place in the code where the local variable has a well-defined value. For example, it is common to test a variable against a limit,

```
x1 = .....;
if(x1 < xmin) x1 = xmin;
else if (x1 > xmax) x1 = xmax;
```

Until the end of the complete sequence, the variable x1 does not have a well-defined value. In many other calculation sequences, a variable might appear several times as various intermediate values are set.

The final step is to prevent any task preemption from occurring while a data transfer is taking place. The simplest way to do this in a uniprocessor system is to disable the processor's interrupts during the transfer,

```
disable_interrupts();
x_x1 = x1;
enable_interrupts();
```

This sequence should be used in any statement that either uses or sets an exchange variable, including within the `get_xxx()` function. The purpose of this is to ensure the integrity of the transferred data by preventing interrupts from occurring in the middle of a transfer. These procedures can only be ignored in those environments that cannot implement preemptive action (such as Matlab).

4.7 Real-Time Realization

Real-time realization is a very small step from simulation. The move to real time can be done in two steps:

- Provide a means of measuring time.

- Substitute an interface to the actual instruments and actuators for the simulation.

The first step produces a real-time simulation. It is still a simulation in that there is no physical system attached, but the program runs on the same time scale as the actual control program. In addition to being useful for debugging and for program development in an environment that precludes connection to the actual system, this mode can also be useful for training. The examples here will be presented in this mode so that the results can be easily duplicated. Some will also be shown in full control mode.

It should again be emphasized that the large amount of attention paid to simulation mode in this chapter is because simulations are extremely valuable to the development process, even if there seems to be a lot of time taken in getting the simulations working. By making code portable between the simulation and the actual control, the methods presented here are intended to minimize the cost of doing a simulation, thereby increasing its attractiveness still further.

Two methods of measuring time will be used, the first of which is *calibrated time*. From a programming point of view, it is exactly the same as simulation. The difference is that the base time increment is adjusted so that when averaged over a sufficiently long period, time as computed in the program matches real time. This method is very easy to use and requires no special hardware. However, it is *not* very accurate and the calibration must be redone for any changes made in the program – even data changes. This is because program or data changes can affect the relative amounts of run time spent in different parts of the program, which will affect the calibration.

The second method uses a hardware clock, that is, a very stable oscillator connected to a logic circuit that can count the oscillator cycles. Several methods will be used to gain access to the hardware clock that is part of most computer systems.

4.8 Real-Time Realization with Matlab

Although Matlab is not intended as a language for real-time realization, with a fast processor it can achieve adequate real-time performance for slow control objects such as thermal or process systems. Where suitable, this would be a very good prototyping environment because of the excellent data display and analysis facilities available in Matlab. The examples here are for implementation in the Microsoft Windows environment. Because Windows multitasking can take control away from

Section 4.8. Real-Time Realization with Matlab

the control program for large time intervals, it is important to make sure no other programs are active at the time the real-time runs are made.

Matlab makes the underlying real-time clock available with calls to the functions `tic` and `toc`. Tic starts a stopwatch type of timer running; toc reads the time since the stopwatch was started (in seconds). Any number of calls to toc can be made to get a running indication of time.

The granularity (minimum measurable increment of time) of the clock can be demonstrated by a short program that keeps reading the time, looking for a change. The relevant file is `timetest.m` in the `grp_mtlb.zip` archive, `templates` directory. When it detects a change the new time is recorded:

```
% Time test loop (script file)
% File name: timetest.m

ttest(1)=0;
i=1;
tic; % tic starts Matlab's stopwatch functions
while i <= 10
    tnew=toc;
    if tnew > ttest(i)
        i = i + 1;
        ttest(i) = tnew;
    end
end
```

Running this gives:

```
ttest =
  Columns 1 through 7
        0    0.0600    0.1100    0.1700    0.2200    0.2800    0.3300
  Columns 8 through 11
   0.3900    0.4400    0.5000    0.5500
```

This shows time change increments alternating between 0.05 and 0.06 seconds; the probable actual increment is 0.055 sec, rounded by Matlab to the nearest 0.01 sec (10ms). This granularity is adequate for the slow control objects for which Matlab implementations would be used.

4.8.1 Heater Control Implementation in Matlab

Experimenting with the heater control program in real-time simulation mode shows that $del\_t = 0.0115$ sec matches real time closely (on a 90MHz Pentium). By replacing one of the output variables with toc a record of real time can be produced to see how good a job the calibrated time is doing. The result of this is shown in figure 4-7 where the ordinate is time as measured by calibration, and the abcissa is the difference between calibrated time and real time. The maximum error shown of about 0.08 sec over 20 sec or running time is actually too optimistic for the Windows environment. A subsequent run of exactly the same program gives the time-error curve shown in figure 4-8. This shows a trend toward an increasing error with an accumulated error of about 0.7 sec in 20 sec of run time.

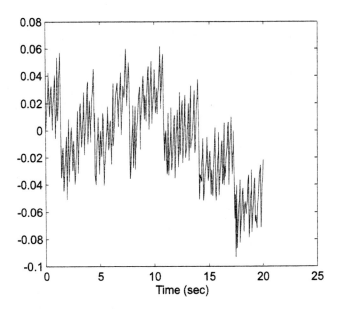

Figure 4-7. Check of calibrated time

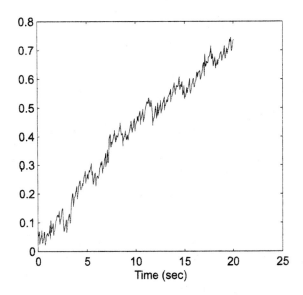

Figure 4-8. Check of calibrated time

Section 4.8. Real-Time Realization with Matlab

Figure 4-7 is useful in looking at the characteristics of the error because of its narrower abcissa scale. The high frequency error appears to be related to the difference in granularity between the two time measures - 0.055 sec for real time and 0.0115 sec for calibrated time. The larger jumps appear to be related to task execution patterns.

Implementing the hardware clock is done by making changes to the `get_time` and `inc_time` functions. The modified functions are shown next. Switching back and forth between simulated and real time involves commenting out the unused portion. These files can be found in the `heatctrl` directory of `grp_mtlb.zip`. The versions of these in the template directory have not been modified.

```
function t=get_time
    % Return the value of running time
    % Can be used for simulation or real time -- comment out appropriate
    % sections for each use

    % Simulation section
    global tstep


    % For real time implementation:
    t = toc;
    tstep = t;

    % For simulated (or calibrated) time
    %t = tstep; % Return simulated time


function inc_time
    % Increments time for simulation use
    % Does nothing for real time use

global tstep del_t % These variables must be declared global

% For real time:
tstep = get_time;

% For simulated (or calibrated) time
%tstep = tstep + del_t;
```

The primary limit in running in this mode is the 55 ms time granularity of the `tic/toc` timer in Windows. With a PWM period of 1 second this yields a resolution of about 20:1. A section of the transition audit trail is shown next. The intervals shown are all divisible evenly by 0.055, demonstrating the timer-limited precision.

```
        13.6700    2.0000    1.0000    2.0000
        13.7800    2.0000    2.0000    3.0000
        14.6600    2.0000    3.0000    1.0000
        14.6600    2.0000    1.0000    2.0000
        15.0500    2.0000    2.0000    3.0000
        15.7100    2.0000    3.0000    1.0000
```

```
15.7100    2.0000    1.0000    2.0000
15.9200    2.0000    2.0000    3.0000
16.7500    2.0000    3.0000    1.0000
16.7500    2.0000    1.0000    2.0000
16.9100    2.0000    2.0000    3.0000
17.7900    2.0000    3.0000    1.0000
17.7900    2.0000    1.0000    2.0000
18.0700    2.0000    2.0000    3.0000
18.8400    2.0000    3.0000    1.0000
18.8900    2.0000    1.0000    2.0000
19.1700    2.0000    2.0000    3.0000
19.8800    2.0000    3.0000    1.0000
```

It is interesting to note that the timing is more accurate when using the hardware timer, but is more precise by almost a factor of 5 when using calibrated time. This suggests the possibility of using some combination of the two, but great care must be taken to make sure that time never goes backward!

Chapter 5

SOFTWARE REALIZATION IN C++

If C or C++ is to be the target language, it makes sense to do the simulation in one of those languages also. Most of the template for direct simulation in C++ is very similar to the Matlab template. Unlike Matlab, some external mathematical library function for approximate integration must be added if there is a continuous-time element of the simulation for the target system. The results must be written to a file since C++ does not have any equivalent to the Matlab "workspace" to leave the result in.

5.1 Simulation in C++

Two software implementations are presented for C++. They differ in how they handle task implementations and to what extent the state transition model is embedded in the base software or left to the user to construct.

The simpler implementation handles groups of tasks, all assumed to have equal priority (this is designated as group-priority in the following sections). Any number of groups can be dealt with to meet priority needs during execution in the real-control system (as opposed to the simulation versions). The group-priority software leaves the construction of most of the state logic to the user. All decisions about which events cause code to execute, when the task should become active, and so forth are also left to the user to write into the code.

The other version provides more services to the user by defining class structures for transition state logic and handling task invocation and priority issues internally. This version is designated TranRun4 in the following sections.

The two versions are presented to show that within the goal of producing portable real-time control software, there are several approaches. The group-priority approach, leaving more to the user but requiring less infrastructure software, is perhaps more applicable to environments with relatively little predefinition, such as new computer boards or new operating systems, and in situations where initial time to market is of prime importance. On the other hand, where the software infrastructure is likely to be reused many times for different applications, there is considerable value in improving programmer productivity by providing a richer infrastructure such as that used in TranRun4.

The general structure of the two implementations is similar: each contains a means of maintaining lists, a task dispatcher for controlling execution of tasks, class definitions for tasks and states, and various utilities for determining time, logging data, and communication. User programs are constructed by instantiating the needed elements and adding code to the member functions associated with those elements.

The reason for constructing these software bases is to insulate the user code from the actual implementation environment. The lowest level of the base software provides the interface to the actual execution environment. It can be changed without changing any of the user-written code. Thus the application design can be tested in a variety of environments or upgraded to new computing hardware with a minimum of conversion effort.

The sample base software is meant to be examined and modified by interested readers. It is all available in source form in the corresonding software archives. The descriptions which follow go one level below the actual user code to give some perspective on how the user's code interacts with the underlying utilities.

A number of sample control systems are used in this and other chapters. In particular, the ficticious manufacturing system, "glue," is used extensively. Chapters 15 and 16 give a step-by-step example of building control software for a simple manufacturing system, beginning with the control of individual components and continuing through the operation of a complete machine. Thus, the sample code snippets that are used here can be related to the larger problem of building an application from scratch.

5.2 Templates for Simulation in C++ (group-priority)

The C++ template makes use of two base classes for building the application program—one for the tasks and one to organize the task lists.[1] These are in the file `basetask.hpp`.

The base class for task definition, `BaseTask`, is quite simple. This is because in the direct program design method used by the group-priority software, all of the properties of the task are embedded into the user-written task code so there is relatively little common information. The most important item in the base class definition is the Run function, which is declared as `virtual`. The actual tasks are defined by inheriting from this base class. Each task must define an actual function of this name which is where all of the state code goes. If the base class Run function is ever called it gives an error message. The other items in the class are the task name, which is used for the audit trail and anything else the user wants to use it for, and some basic data that all tasks need. There are also elements and access functions for intertask communication.

---

[1] archive: grp_cpp.zip, directories: template, common

Section 5.2. Templates for Simulation in C++ (group-priority)

```cpp
// Base class for task template.  The class must have a public Run() function.
class BaseTask
    {
    public:
        BaseTask(char *aName,int ProcNo = 0);   // Constructor
        void BaseTaskInit(char *aName,int taskcode); // Initialization
        virtual ~BaseTask();
        void ActivateProfiling(double vbegin,double vend,
                int num_bins,int loglin);
        virtual int Run(void);
        long GetNumScan(void);
        int GetTaskProcess(void){return ProcessNo;}
        int GetTaskCode(void){return TaskCode;}
        void StartProfile(void);  // Do the profile calculations
        void EndProfile(void);
#ifndef VC_RTX_RTSS  // RTSS has no file system
        void WriteProfile(ofstream &OutputFile);
#endif  // VC_RTX_RTSS  // RTSS has no file system

        void SendRTMessage(BaseTask *ToTask,int BoxNo,int Flag = 1,
                double Val = 0.0,int mode = 0,int ReturnReceipt = -1);
        void SendRTMessage(int ToTask,int BoxNo,int Flag = 1,
                double Val = 0.0,int mode = 0,int ReturnReceipt = -1);
        int GetRTMessage(int BoxNo,int *msg_flag = NULL,
            double *val = NULL,int *from_task = NULL,int *from_process = NULL);
        // Delivery modes:
        //  0  Only deliver message if message box is empty (ie, flag=0)
        //  1  Deliver message regardless of message box state
        char *Name;    // Task name
        int TaskID;    // ID number that can be used to track tasks
        int TaskCode;  // Will be set to same as TaskID unless explicitly
                       // set
        double GetGlobalData(BaseTask *pb,int k);
        int AcceptMessage(int BoxNo,int msg_flag,double msg_val,int mode,
                int from_task,int from_process);  // This was private,
                // but is has been moved to public so it can be
                // accessed by StoreMessage() and other network
                // interface functions. They could have been declared
                // to be 'friend' functions, but that would make the
                // placement of #define USE_UDP very critical.
        void SendCompletionMessage(int toTask,int &respBox,int code);
        int AcceptGlobal(double *pData,int nData);
        double *GetGlobalArray(BaseTask *pb,int *n);
        void CreateStateNameArray(int n);
        void RegisterStateName(char *pName,int i);
        char **GetStateNames(void);

    private:
        Message_Box mbox[NUM_MESSAGE_BOX]; // This is private so
           // it can't be accessed by derived classes -- this
           // protects the integrity of the message
        void SendMsg(BaseTask *ToTask,int BoxNo,int Flag,
                double Val,int mode,int ReturnReceipt); // Internal function
        double *GlobalData;  // Global information for this task -- accessible
                // to all other tasks via call to GetGobalData()
```

```
protected:
    int State, xState;      // Keep track of the current State.
    int NextState;  // The State to be run next execution scan. -1 means no change
    int RunEntry;   // Flag set on State change to cause entry function to run
    double NextTaskTime, xNextTaskTime;// The next time this task should run
    double DeltaTaskTime, xDeltaTaskTime;// Period for this task
    long NumScans;  // For performance tracking
    int ProcessNo;  // Process that this task is located in
    // Data associated with task profiling
    int profile_on; // Flag to determine if profiling is on
    int nbin;  // number of bins
    double *values;  // Boundary values for histogram bins
    double v1,vlast;  // Low and high boundaries for histogram
    long *occur;  // Record of occurrences by bin
    double Time0;  // Time at start of task scan
    double *PrivateGlobalData;  // Global data array -- must be copied
               // to public version for use by other tasks
    int GlobalSize;  // Number of global variables
    int CopyGlobal(void);  // Copy the private global variables to
      // public view -- for multiprocess systems this copies the
      // private global set to all processes
    int CreateGlobal(int nGlobal);  // Create global arrays
    char **StateNames;  // Array of pointers to store state names
    int nStateNames;
};
```

A functioning control program consists of a set of tasks that has been instantiated, with user-written code inserted into the Run() function. Lists of these tasks are maintained in order to access them for execution or for information passing. The following classes define the classes that are used for these lists.

Two classes are defined for list maintenance, TListElement and TList. The first is used to construct a list element, and the second to construct a list. These provide for construction of a linked list of items associated with the BaseTask class, which will be used for scheduling. They are:[2]

```
// TListElement - data associated with any element on a list
class TListElement
    {
    public:
        TListElement(BaseTask *ItemPointer);
        ~TListElement(void);
        TListElement *Next;
        BaseTask *Item;
    };

// TList - a list of BaseTasks.  The list allocates space
// for itself as elements are added via Append.
class TList
```

---

[2] The reader may wonder why container objects available in the Standard Template Library are not used. This software was first written before compilers supporting the STL became widely available. Future versions will make use of standard container classes.

Section 5.2. Templates for Simulation in C++ (group-priority)

```
{
public:
TList (void);
~TList (void);

void Append (BaseTask *Item);     // Add Item to end of list

BaseTask *NextItem(void);         // Return pointer to next
                                  // item on list

int IsEmpty(void);                // Is the list empty
int HowLong(void);                // How many items on list

private:
TListElement *First;              // First element on list
TListElement *Last;               // Last element on list
TListElement *Current;            // Last element on list
int Length;                       // Number of elements on list
};
```

The file `basetask.cpp` contains the functions associated with these classes. The only task-related function is the default Run function, which prints an error message. The other functions are for lists, adding and removing items, and so forth.

The complete main file is included here. It follows the format of the Matlab version fairly closely with one significant difference: the simulation is handled as a separate task rather than built into the main execution loop.

The **new** operator is used for most of the item creation activities. In the initialization part of the main function, **new** is used to set up the list of tasks, and also to set up the arrays for the output information.

The execution loop has two major sections. It first checks to see if outputs should be added to the output array and then scans the list of tasks using the functions from the list processing class. Finally, after the execution loop is done, the outputs from the output data and the audit trail are written to data files.

```
// Template for the "main" file of a group priority application
// File: mainfile.cpp
// Created by D. M. Auslander and J. Jones, 12/19/95
// (based on existing examples)
// This file shows skeletons of all of the component
// parts of an application. It cannot be compiled because
// the actual code is missing. It can be used to build a
// program by filling in the missing pieces and changing
// names (away from the generic names such as "task1")

// There are also template files for the task file and for
// the header file, tasks.hpp

#define MAIN_CPP

#include <stdio.h>
#include <stdlib.h>
#include <conio.h>
```

```
#include <math.h>
#include <dos.h>
#include "tasks.hpp"        // Task and environment info

static StopControlNow = 0;  // Variables and function used to
char *StopMessage = NULL;   // turn the program off
void StopControl (char *aMessage)
    {
    StopControlNow = 1;
    StopMessage = new char[strlen(aMessage+1)];
    strcpy (StopMessage, aMessage);
    }

double EndTime = 1.0e10; // make this value visible to other function
                         // in this file (use big number to run forever)

void main(void)
    {
    double TheTime;         // Running time

    // The task lists are instantiated here
    // Declaration of TList pointers are in tasks.hpp
    LowPriority = new TList("Low Priority");
    Intermittent = new TList("Intermittent");
    #ifdef ISR_SCHEDULING
    // These lists are only instantiated if interrupt
    // scheduling is used.
    Preemptable = new TList("Preemptable");
    Interrupt = new TList("Interrupt");
    #endif // ISR_SCHEDULING

    // Create all of the task objects used in the project
    Task1 = new CTask1("Task1");
    Task2 = new CTask2("Task2");
    ...                    // Other tasks

    // Instantiate the task that handles the operator interface
    // (if any)
    OpInt = new OperatorInterface("Oper Int");
    // Define another task object for running the simulation.
    // This task is scheduled just as any other tasks, and is
    // only added to the task list if a simulation is desired.
    // The actual simulation code in contained in the Run()
    // member function of SimulationTask.
    // Note that this use if the #define SIMULATION is optional.
    #ifdef SIMULATION
    SimTask = new SimulationTask("Simulation");
    #endif //SIMULATION

    // All tasks must be added to the task list here.
    #ifdef ISR_SCHEDULING
    Preemptable->Append(Task1);
    Interrupt->Append(Task2);
    #else // ISR_SCHEDULING
    Intermittent->Append(Task1);
```

Section 5.2. Templates for Simulation in C++ (group-priority)

```cpp
    LowPriority->Append(Task2);
#endif // ISR_SCHEDULING

    LowPriority->Append(Task3);
    LowPriority->Append(OpInt);

#ifdef SIMULATION
    LowPriority->Append(SimTask);
#endif //SIMULATION

    // Set up a data output object
    // The arguments for the DataOutput constructor are:
    //    int nvars,double STime,double LTime,double DTime
    DataOutput *DataOut1 = new DataOutput(3,0.0,20.0,0.1);

    // You must call this function to initialize the Audit
    // trail array before you write to it.
    InitAuditTrail();

    // The delay to maximize the window is necessary if the
    // operator interface is being used in an environment that
    // uses sizable windows (e.g., EasyWin)
    cout <<
        "Maximize window if necessary. ENTER to continue...\n";
    getch();
    ProjectSetup(); // Get initial parameter values

#ifdef ISR_SCHEDULING
    SetupTimerInterrupt();
#else // ISR_SCHEDULING
    // Set the proper mode for the timer
    SetupClock(-1.0);
#endif // ISR_SCHEDULING

    // Scheduler loop.  Inside this loop, you MUST NOT call
    // any terminal I/O functions such as printf, cout,
    // getchar, scanf, etc. (they are either too slow or are
    // blocking).
    //   This includes inside the run
    // functions of the tasks.  These things are OK in
    // SIMULATION ONLY.  Use #ifdef SIMULATION (or something
    // similar) if you have sections of code in which you
    // want to print to the screen (for debugging).
    while (((TheTime = GetTimeNow()) <= EndTime)
            && (TheTime >= 0.0) && (!StopControlNow))
        {
        // This section is used for general output. The DataOutput
        // object store data in arrays until requested to write the
        // data to the disk. This could be done is a separate task,
        // if desired.
        if(DataOut1->IsTimeForOutput())
            {
            // Get data for output
            double val,mc,set;
```

```
            Task1->GetData(&val,&mc,&set);
            // Data from the controller
            DataOut1->AddData(Task2->GetSetpoint(),
                            val, mc, END_OF_ARGS);
            IncrementTime();
          // Account for time spend in output recording
            }
        // Run LowPriority and Intermittent list
        if (TaskDispatcher(LowPriority,Intermittent))
           StopControl("\n***Problem with background tasks***\n");
        }
#ifdef ISR_SCHEDULING
RestoreTimerInterrupt();

#endif // ISR_SCHEDULING

clrscr (); // Clear the operator interface

if (StopMessage == NULL)
     cout << "\n\n***Normal Program Termination***\n\n";
else
     cout << StopMessage;

// There are two versions of the audit trail, one with task
//   names and one with task numbers instead (for Matlab)
WriteAuditTrail("trail.txt");
WriteAuditTrailNum("trl_num.txt");

cout << "\nThe time at termination is " << TheTime << "\n";
// Print the scan statistics to the screen
WriteScanStat("scans.txt", TheTime);
// Save the output data to a file
DataOut1->WriteOutputFile("data_out.txt");
// If any tasks have created their own output records,
// the files can be written here
Task2->WriteEventRecord("evnt_rec.txt");
// De-allocate space give to tasks
delete (LowPriority);
delete (Intermittent);
#ifdef ISR_SCHEDULING
delete (Preemptable);
delete (Interrupt);
#endif // ISR_SCHEDULING
}
```

This template includes several methods of dispatching (or scheduling as it is referred to in the internal comments). For the examples in this chapter, only purely sequential dispatching is used. This is accomplished with the LowPriority list; all other lists are left empty.

The execution loop, which is in the main file in the Matlab version, is in the function TaskDispatcher(), which is in the file basetask.cpp. The execution loop itself from this function is shown below. Only the part of the loop that processes the LowPriority list is shown:

Section 5.2. Templates for Simulation in C++ (group-priority)

```
...
for (int j=0; j < jmax; j++)
    {
...
// Stuff goes here associated with the other list
// This section processes the LowPriority list
    if(List1Num == 0)break; // No items in List1
    if ((TempTask = List1->NextItem()) == NULL)
        {// Get the next Task on the list
        cout << "\nTried to access empty task list\n";
        return(-1);
        }
    Suspend = TempTask->Run(); // Run the Task
    IncrementTime();
    if (Suspend == -1)
        {
        cout << "\nProblem running Task: " <<
                TempTask->Name << "\n";
        return(-1);
        }
    }
...
```

This code cycles through all of the tasks in the list once. Each task gets one scan, that is, one call to its **Run** function. The process of defining the tasks starts with the definition of a class for each task. These go into the header file, `tasks.hpp` (by convention, the name `tasks.hpp` is used for all projects).

An important distinction must be observed between task classes and task objects. Like any other class, a task class is a description of how to construct an object. No object actually exists until it is instantiated. A task (i.e., an object) is created when it is instantiated from a task class (usually with the operator **new**). In fact, any number of tasks can be created (instantiated) from a single task class. They would all share the same general structure, but each would have its own unique data.

Task classes are derived from the **BaseTask** class. A typical class looks like:

```
class CTask1 : public BaseTask
    {
    public:
        Task1(char *aName);      // Constuctor for the class.
        ~Task1();       // Destructor for the class.
        int Run(void);
        // Run() is the function containing the actual task code
        // that the scheduler loop calls. The return value can
        // be used as desired.
        // These are the functions that allow other tasks to
        // access this one
        void SetSomething(double);
        ...
        double GetSomething(void);
        ...
```

```
        // Other functions, this one for example is a "cleanup"
        // function that is used to write a data file after
        // the real time operation is done
        void WriteDataRecord(char *filename); // Write a data file
    private:
        // This section is for variables that are private to
        // this class. The local variables and the "exchange"
        // variables are placed here (the convention is to
        // use an 'x' as the first letter of the exchange variable
        // names).
        double A_variable,xA_variable;
        // The 'x' indicates an exchange variable
        double Another_variable,xAnother_variable;
        ...
        // Functions that are private to this class
        void DoSomething(double);
        ...
        // This variable is used to keep track of data internal to
        // this class
        DataOutput *EventRecord;
    };
```

Of critical importance in this definition is the private part in which the variables associated with the particular task type are defined. This is a departure from the Matlab model in which all of the variables are put together into a global group. Here the variables are isolated into the class for the task, and the variable initialization takes place in the constructor function. The C method of putting all variables as static in the task file is similar to this, but much less formalized and more prone to error.

Several **#define** definitions are included in this header file to define the computational environment and pointers to the task instances. These are necessary so that all tasks can access the public functions of other tasks, most commonly to transfer information. These declarations are shown here; the separate declaration for the main file is so that global storage is allocated there.

Tasks need to be visible to all other tasks. To accomplish that, pointers to the instantiated tasks are maintained as **extern** variables. The definition for EXTERNAL provides for the proper C++ syntax of the use of the **extern** keyword, so that memory is only assigned once. The sample is from the glue02.cpp sample program.

```
#ifdef MAIN_CPP
    #define EXTERNAL  // So externs get declared properly
#else
    #define EXTERNAL extern
#endif

// Tasks
EXTERNAL CDataLogger *DataLogger;
EXTERNAL PIDControl *BeltControl1,*BeltControl2;
EXTERNAL PIDControl *HeatControl1,*HeatControl2;
```

Section 5.2. Templates for Simulation in C++ (group-priority)

```
#ifdef SIMULATION
    EXTERNAL CBeltAxis *BeltAxis1,*BeltAxis2;
    EXTERNAL CThOven *ThOven1,*ThOven2;
#endif
```

The code associated with each task class is collected in a file, one file (or more) per task. The functions in this file include the constructor, where everything relevant to this task class is initialized, the destructor, which is empty in most cases (it is needed if any memory was allocated in the constructor), and the Run function that defines the state-transition structure and associated code. The template for a task file is `taskfile.cpp`:

```
// Template file for group priority applications
// This file gives the skeleton of a typical task file
// In most applications there is a file like this for
// each task.
// File: taskfile.cpp
// Created by DM Auslander and J Jones, from other examples

#include <iostream.h>
#include <math.h>
#include "tasks.hpp"

CTask1::CTask1(char *aName)     // Constructor for the task class
        :BaseTask(aName)
    {
    // These variables come from BaseTask so must be specified
    // for all tasks
    DeltaTaskTime = 0.1; // Task period
    NextTaskTime = 0.0;

    State = 0;        // set to initialize
    NextState = 0;
    // Data specific to this task class
    // If more than one instantiation (i.e., more than one task)
    // is going to be created from this class,
    // data that is different across
    // instantiations can be passed in as arguments to the constructor
    ...
    }

CTask1::~CTask1(void){};  // Destructor (often empty)

// Member functions for the class CTask1
// Most classes will have GetSomething() and SetSomething() functions
// for intertask communication. There could be other member functions
// as well

double CTask1::GetSomething(void)
    {
    double rv;
    CriticalSectionBegin(); // Protect the intertask transfer
    rv = xvalue;         // "x" indicating and exchange variable
    CriticalSectionEnd();
    return(rv);
    }
```

```cpp
// Other member functions ....

// The 'Run' function
int CTask1::Run(void)    // Every task class must have a "Run" function
    {
    double Time;
    int done = 1;        // Default return value indicating
                         // done for this event
    NumScans++;          // Increment scan count

    if (State == 0)
        {
        // Initialization code.  If there is anything
        // that the task needs to do only once at
        // start-up, put it here.
        State = 1;
        NextState = 1;
        return (0);
        }

    // Only run the task if it is proper time.
    // Note:  It is not required that there be some time-
    // based criteria for when the task runs.  This
    // section could be taken out, resulting in a task
    // that runs every scan. That is a design decision
    // based on the nature of the task.

    if ((Time = GetTimeNow()) < NextTaskTime)
        return (done);    //  Done for now

    // Calculate the next run time for the task
    NextTaskTime += DeltaTaskTime;

    if (NextState != -1)
        {
        // Get here if there has been a transition, indicated
        // by a nondefault value of NextState

        // record audit trail here if desired

        AuditTrail(this, State, NextState);
        State = NextState;
        NextState = -1;
        RunEntry = 1;
        }
    else
        RunEntry = 0;

    // This is the state transition section of the code
    // The variable "State" is maintained in BaseTask,
    // the parent class for all task classes
    switch (State)
        {
        case 1: // State1 -- indicate state name here for easy
                // identification

            if (RunEntry)
```

Section 5.2. Templates for Simulation in C++ (group-priority)

```
                {
                // only run this on entry to the state
                // entry code goes here
                ...
                // It is common to exchange information in this
                // section which needs a critical section,
                CriticalSectionBegin();
                xsomething = something;
                CriticalSectionEnd();
                ...
                }
            // Action Section - is run on every scan (often empty)
            ...

    // Test/Exit section
    if(--test something--)
        {
        NextState = 2;      // Go into another state
        done = 0;           // Run next state immediately
        }
    break;

    case 2: // State2
        // Entry Section
        if(RunEntry)
            {
            ...
            }
        // Action Section
        ...

        // Test/Exit Section
        if(--test something--)
            {
            NextState = 3;  // A transition
            done = 0;       // Go to next state immediately
            }
        else if(--test for something else--)
            {
            NextState = 6;  // A transition
            done = 1;       // Only go to the next state after
                            // other tasks have been serviced
            }

        break;

        ...             // Additional states...

        default:        // check for an illegal state.

        cout << "\nIllegal State assignment, Task1\n";
        return (-1);
    }
    return (done);
}
```

This template includes code that only runs the function at specified time intervals. This code is optional. It is used to improve efficiency by not actually entering any of the state code if it is known in advance that the task does not need to run.

5.3 PWM Simulation Using C++ (group-priority)

The task that does the PWM generation follows the template outline.[3] Its class is derived from the BaseTask class as follows (in tasks.hpp; remember that the file name tasks.hpp is reused in all of the project directories):

```
class PWMTask : public BaseTask
    {
    public:
        PWMTask(char *aName);       // Constuctor for the class.
        ~PWMTask();                 // Destructor for the class.
        int Run(void);
        // Run() is the function containing the actual task code
        //  that the scheduler loop calls. The return value can
        //  be used as desired.
        void SetPeriod(double);
        void SetDutyCycle(double);
        double GetPWMOutput(void);
        void WritePWMRecord(char *filename);  // Write a data file
    private:
        double DutyCycle,xDutyCycle;
        // The 'x' indicates an exchange variable
        double PWMOutput,xPWMOutput;
        // Variable used for simulating output
        double EndOnTime;
        double EndOffTime;
        double Period,xPeriod;
        void PWMOn(void);
        void PWMOff(void);
        DataOutput *PWMRecord;
        // Output object to record output changes
    };
```

All of the variables associated with PWM generation are declared in this class as are the public functions that allow other tasks to control the PWM (SetPeriod() and SetDutyCycle()). The functions PWMOn and PWMOff are private because they are used internally by the PWM task to turn the output on or off. There is no need for public access to these functions.

The Run() function, which is declared as virtual in the parent class, is where the state-related code resides. A section of this is (from pwm1.cpp):

```
switch (State)
    {
    case 1:      // State 1: "Compute_Times"
        // Entry Section
        if (RunEntry)
            {
```

---

[3]archive: grp_cpp.zip, directory: pwm1

Section 5.3. PWM Simulation Using C++ (group-priority)

```
            // only run this on entry to the state
            // entry code goes here
            // Copy the latest value for duty cycle and period

            CriticalSectionBegin();
            DutyCycle = xDutyCycle;
            Period = xPeriod;
            CriticalSectionEnd();
            // limit the value of DutyCycle to [0,1]
            if(DutyCycle < 0.0)
                DutyCycle = 0.0;
            else if(DutyCycle > 1.0)
                DutyCycle = 1.0;
            EndOnTime = DutyCycle*Period+EndOffTime;
            EndOffTime += Period;
            }
        // Action Section
        // no action code for this state
        // Test/Exit section
        // transition tests go here.
        if(DutyCycle > 0.0)
            NextState = 2;
        // exit code goes here
        // No exit section for this transition
        else if (DutyCycle <= 0.0)
            NextState = 3;
        // exit code goes here
        // No exit section for this transition
        done = 0; // Continue in this task
        break;
    case 2: // State 2: "PWM_On"
        // Entry Section
        if (RunEntry)
            {
            // only run this on entry to the state
            // entry code goes here
            // turn the PWM output on
            PWMOn();
            }
        // Action Section
        // no action code for this state
        // Test/Exit section
        // transition tests go here.
        if((Time >= EndOnTime) && (DutyCycle < 1.0))
            NextState = 3;
        // exit code goes here
        // No exit section for this transition
        else if((Time >= EndOnTime) && (DutyCycle >= 1.0))
            NextState = 1;
        // exit code goes here
        // No exit section for this transition
        break;
    case 3: // State 3: "PWM_Off"
        // Entry Section
        ...
```

The intialization and instantiation takes place in the main file, pwm1test.cpp. This follows the pattern shown in the previous template:

```
...
// Create all of the task objects used in the project
PWM = new PWMTask("PWM");
PWMSup = new PWMSupTask("PWMSup",0.4);
// 2nd argument is the ...
...
```

where PWM and PWMSup are pointers to the tasks based on the classes declared for each task. The rest of the file is almost exactly the same as the template.

5.4 Simulation in C++ (with TranRun4)

The TranRun4 scheduler is designed to provide the same functionality as the group-priority software while insulating the user from many of the details of the implementation. The group-priority software places most of the code which controls task timing and state transition logic together with the user's code. In contrast, the TranRun4 scheduler provides a set of utility files which contain the code necessary to implement a real-time program. The user interacts with the utility code using a formalized Application Programming Interface (API) that consists of function calls and C++ objects. The user does not need to make any changes to the utility files in order to create all the sorts of real-time programs for which the scheduler is designed. However, the source code is freely available and the reader is encouraged to study, modify, extend, and improve the code as needed.

In order to create a TranRun4 program, the user must create a set of task and state objects and configure the scheduler to run these tasks and states in the desired fashion. Task and state objects are created by the method of inheritance. All the code which controls task timing and state transitions is provided in the base classes CTask and CState. The user derives classes from CTask and CState, instantiates objects of these derived classes, and configures the scheduler to run the objects. The code which the user needs to run as entry, action, and transition test code is placed in member functions belonging to the user's derived task and state classes.

The traditional arrangement of files in a TranRun4 project has one header file (*.hpp or *.h) and one source file (*.cpp) for each task. There is an additional pair of files designated the main header and source file. The main source file contains the UserMain() function, which is the user's entry point function to the program.

5.4.1 Components

The software model described in this text breaks down control systems into sets of tasks and states; software is organized into processes and threads. The components of TranRun4 are organized as a class hierarchy which follows the task, state, process, and thread model closely. Additional classes are added to provide functionality associated with such needs as timing and list maintenance. Table 5.1 lists the most

Table 5.1. Major classes in TranRun4

Name	Files	Function
CMaster	TR4_Mstr	Main scheduler object which oversees all the others
CProcess	TR4_Proc	Controls a set of tasks which run on one computer
CTask	TR4_Task	Base class for all task objects
CState	TR4_Stat	Base class for all state objects
CMessageInbox	TR4_Shar	Holds data which has been sent by other tasks
CGlobalData	TR4_Shar	Holds data to be shared with all other tasks
CRealTimer	TR4_Timr	Keeps track of real or simulated time
CProfiler	TR4_Prof	Measures the execution time of the user's functions
CDataLogger	Data_Lgr	Saves data in real time, writes to a file later

important classes with which the user interacts; there are a number of other internal classes with which users normally do not have to deal. The file names given are the base names; there will be one header (*.hpp) file and one source (*.cpp) file for each file name.

Classes CMaster, CProcess, CTask, and CState form a functional hierarchy which controls the scheduling process. There is exactly one object of class CMaster instantiated for every real process [4] in which the scheduler runs. There may be one or more objects of class CProcess attached to the master scheduler. The presence of multiple process objects is allowed in order to provide a framework in which to run multi-process systems—systems in which several executable programs are running on the same computer or even on different computers. Each process object maintains a set of lists of tasks which collectively contain all the task objects which exist within a given project. Each task object maintains a list of states, and each state has a set of entry, action, and transition test functions to run. The primary goal of the whole system is to ensure that those entry, action, and test functions run when they are supposed to.

When creating a simulation program using TranRun4, the user creates a set of processes, tasks, and states. The processes (usually just one) are then associated with the master scheduler object, the master scheduler having been created automatically within the base software. The tasks are attached to their process and the states are attached to their tasks. Then when the master scheduler is run, the flow of control follows the same pattern: The master runs a simple while loop. During every step through the loop, the master tells the active processes to run. These processes tell their tasks to run, and the tasks tell the appropriate states to run. The state objects then call the entry, action, and transition test functions which have been written by the user.

---

[4] The distinction between process objects and real processes is as follows: *real processes* are the independent physical processes running on one or more computers, each having its own address space; *process objects* are objects of class CProcess which represent real processes but may, for simulation and debugging purposes, be running within the same real process.

5.4.2 The Master Scheduler

The master scheduler is primarily responsible for maintaining a list of processes and running the main scheduling loop. The main scheduling loop works in just the same way as the loop in the simplest possible simulation program: While the simulation is unfinished, it calls a set of functions once for each time step. The code which performs this function, contained in `CMaster`'s method `Go()`, is shown next. Some code which handles special hardware and software has been removed for clarity.

```
// Now run the background tasks by repeatedly calling RunBackground() and any
// other functions which need to be run in the given real-time mode. Exit the
// loop when time is StopTime or someone called Stop() to set Status to not GOING
while (Status == GOING)
    {
    // Call function to cause all nonpreemptively scheduled tasks to run
    RunBackground ();
    ...
    // Check the time; if time's up, say so, stop timer, and cause an exit
    if ((TheTime = GetTimeNowUnprotected ()) > StopTime)
        {
        Status = STOPPED;
        TheTimer->Stop ();
        *ExitMessage = "Normal scheduler exit at end time ";
        *ExitMessage << (double)TheTime << "\n";
        }
    ...
    }
```

The `Go()` method also starts and stops the timer object and handles synchronization between processes.

Tasks are actually scanned during the call to `RunBackground()`. The definition of background and foreground execution used in this text is the opposite of the definition often used by the programmers of non-real-time software: here, background tasks are lower priority tasks which are scheduled cooperatively, while foreground tasks are preemptively scheduled high priority tasks. In simulation mode, all tasks are run in a single thread without preemption, so all tasks are run in the background. The `RunBackground()` function simply calls an identically named function which belongs to the process objects as follows:

```
void CMaster::RunBackground (void)
    {
    CProcess* pProcess = (CProcess*) GetHead ();
    while (pProcess != NULL)
        {
        pProcess->RunBackground ();
        GetObjWith (pProcess);
        pProcess = (CProcess*) GetNext ();
        }
    }
```

The function `GetHead()` retrieves a pointer to the first object in the master scheduler's list of processes. `GetNext()` retrieves a pointer to the next element in the list. These functions are provided as part of class `CListObj`, an ancestor class of `CMaster` which is defined in file `Base_Obj.hpp`.

The master scheduler class also contains a number of utility functions which can help the user with analyzing and debugging software. Also present are functions which allow global scheduler settings to be changed. The most important of these functions will be introduced in the description of the `UserMain()` function which follows.

5.4.3 Process Objects and Task Lists

The principal responsibilities of a project object are to maintain a set of task lists and to run tasks from those lists when asked to by the master scheduler. Various utility functions are provided by the process as well. An abbreviated version of the `CProcess` class definition follows.

```
class CProcess
    {
    private:
        char *Name;                              // Name of this process
        CTimerIntTaskList *TimerIntTasks;        // Unitary interrupt tasks
        CPreemptiveTaskList *PreemptibleTasks;   // Preemptible interrupt tasks
        CPreemptiveTaskList *BackgroundTasks;    // Sample time and digital event tasks
        CContinuousTaskList *ContinuousTasks;    // List of continuous tasks
        ...

    public:
        CProcess (const char*);                  // Constructor which we use sets name
        ~CProcess (void);                        // Virtual destructor called by list

        void WriteStatus (const char*);          // Dump status of all tasks
        void WriteConfiguration (const char*);   // Dump task configuration information
        void ProfileOn (void);                   // Turn profiling on for all tasks
        void ProfileOff (void);                  // Turn all tasks' profiling off
        void WriteProfiles (const char*);        // Dump info about how fast tasks ran
        void WriteProfiles (FILE*);              // Same function as called by CMaster
        void RunBackground (void);               // Run one sweep through task lists
        void RunForeground (void);               // Run a sweep through ISR driven list

        // The AddTask() methods create a new task object and add it into the task
        // list for this process, returning a pointer to the newly created task.
        // InsertTask() is called to put an already-created task into the task list.
        CTask *AddTask (const char*, TaskType);
        CTask *AddTask (const char*, TaskType, int);
        CTask *AddTask (const char*, TaskType, real_time);
        CTask *AddTask (const char*, TaskType, int, real_time);
        CTask *InsertTask (CTask*);
        CTask *InsertTask (CTask&);
    };
```

The four task lists hold pointers to tasks which will be scheduled in four different modes. For purposes of simulation, we are only interested in one mode, "continuous" mode; the other modes will be discussed in chapter 8. The functions `AddTask()` and

InsertTask() place tasks into a task list; their use is discussed in the description of the UserMain() function. The tasks are scanned during calls to RunBackground() and RunForeground(). In simulation, of course, only RunBackground() is used. The portion of RunForeground() which is used in the simplest simulation mode is:

```
void CProcess::RunBackground (void)
    {
    //  If in single-thread, sequential simulation mode: just run all tasks in order
    #if defined (TR_EXEC_SEQ) && defined (TR_THREAD_SINGLE)
        TimerIntTasks->RunAll ();
        PreemptibleTasks->RunAll ();
        BackgroundTasks->RunAll ();
        ContinuousTasks->RunAll ();
    #endif
    ...
    }
```

Thus for a simple simulation program, the process object simply calls the RunAll() function belonging to the list of continuous tasks. That function is:

```
bool CTaskList::RunAll (void)
    {
    CTask* pCurTask;                    //  Pointer to the task currently being run

    //  Send a run message to each task in the list, in sequence
    pCurTask = (CTask*) GetHead ();
    while (pCurTask != NULL)
        {
        pCurTask->Schedule ();
        pCurTask = (CTask*) GetNext ();
        }
    return true;
    }
```

Finally we can see where the tasks are caused to do something: the function Schedule() belonging to each task is called.

5.4.4 Task Objects

Class CTask is the most complex class in the TranRun4 system. It is the base class for all task classes, so it contains the functionality which is common to all task objects. In real time or in simulation, timing is controlled at the task level—each task runs on its own timetable, or in response to its own triggering event. A highly abbreviated copy of the CTask class definition is shown next. Many utility functions and internal variables have been removed so that we can concentrate on the most important functions of the task—maintaining a state list and controlling its timing.

```
class CTask : public CBasicList
    {
    private:
        real_time TimeInterval;         //  Interval between runs of task function
        real_time NextTime;             //  Next time at which task func. will run
```

Section 5.4. Simulation in C++ (with TranRun4)

```
        CState* pInitialState;           // First state to be run by scheduler
        CState* pCurrentState;           // State currently being run by this task
        ...

    protected:
        long State;                      // The TL state in which this task is now
    public:
        // There are lots of constructors for different task types and one destructor
        CTask (CProcess*, const char*, TaskType);
        CTask (CProcess*, const char*, TaskType, real_time);
        CTask (CProcess*, const char*, TaskType, int);
        CTask (CProcess*, const char*, TaskType, int, real_time);
        virtual ~CTask (void);

        // For state-based scheduling, Run() must be overridden by the user; for task
        // based scheduling it calls states' Entry(), Action(), and TransitionTest()
        virtual void Run (void);

        TaskStatus Schedule (void);      // Decide which state functions to run
        void SetSampleTime (real_time);  // Function resets interval between runs
        real_time GetSampleTime (void)   // Function allows anyone to find out the
            { return (TimeInterval); }   //    sample time of this task
        void InsertState (CState*);      // Insert a new state into the state list
        void SetInitialState (CState*);  // Specify the state in which we start up
        ...
    };
```

The variables `TimeInterval` and `NextTime` control when a task will run. The assumption made is that in many situations, a task will not need to run every time when the processor has time in which to run the task. This assumption is closely tied to the definition of intermittent tasks, discussed in section 8.1.1. Therefore a task can be configured so that it only runs when the time (be it simulated or real time) has reached a certain point, or when some external event has occurred. This feature can often be of use when in simulation mode, as the relative frequencies of task scans can be varied. For example, one might wish to study the effect of the frequency at which a controller is run on the controller's performance. The simulation and controller tasks can be configured so that the simulation task runs at a high frequency to accurately model continuous-time behavior, while the controller task runs at a much lower frequency. The frequency of task runs, in real or simulated seconds, is set with a call to `SetSampleTime()`. The argument is of type `real_time`, a type which is set by a `typedef` statement to be equal to `double` in most implementations of TranRun4.

Pointers are present which allow a task to keep track of its initial and its current states. These pointers refer to objects of classes which are descendents of class `CState`. Every task must have an initial state, which must be set by a call to `SetInitialState()`. The argument to this function is a pointer to a state object; the creation of state objects is discussed later. The current state pointer is used to implement state transitions in real time based on the return values of `TransitionTest()` functions which belong to state objects. The `InsertState` function is used to insert a state pointer into the task's list of state pointers. This requires, of course, that a valid state object has been created already. A task can have any number of states, implementing any valid state transition diagram.

There are several task constructors. These are used to create tasks of different scheduling modes. The need for tasks which operate in different scheduling modes (such as continuous and intermittent is discussed in chapter 8. For most basic simulations, all tasks will be of the continuous type; when timing tests are being done, intermittent tasks may be used.

In order to scan a task, the task's parent process calls the task's member function Schedule(). Inside this function is code which determines if the task is ready to run at the current time. For example, if a task has been configured to run every 10 milliseconds last ran 2 milliseconds ago, the task will refuse to run at this time; it will return control to the scheduler so that other tasks can run instead. If it's time for the task to run, Schedule() calls another member function called Run(). Inside Run() is code which finds the current state and calls the Schedule() method belonging to that state.

```
void CTask::Run (void)
    {
    CState* pNextState;                   // Pointer to the next state to be run

    // If this is very first time to run (or restart), go to specified initial state
    if (FirstTimeRun == true)
        {
        FirstTimeRun = false;             // Well it's not gonna be true next time

        // If no initial state has been specified, we can't run; so exit
        if (pInitialState == NULL)
            TR_Exit ("No initial state specified for task \"%s\"", Name);
        // Else, an initial state was given, so go to its location in the list
        else
            pCurrentState = pInitialState;
        }

    // Make sure there really is a state to run; if not, complain and exit
    if (pCurrentState == NULL)
        {
        TR_Exit ("No states or Run() function specified for task \"%s\"", Name);
        }
    else
        {
        // The CState::Schedule() method returns a pointer to the next state, or
        // NULL if no transition is going to occur this time
        if ((pNextState = pCurrentState->Schedule ()) != NULL)
            {
            // If transition logic tracing is enabled, write a line in the trace file
            if (Do_TL_Trace == true)
                TL_TraceLine (this, pCurrentState, pNextState);

            pCurrentState = pNextState;
            }
        }
    }
```

If you've followed the trail of function calls from the master scheduler's execution loop to here (five levels deep and counting), you may be wondering why there are separate `Schedule()` and `Run()` functions belonging to each task. The answer is flexibility—sometimes the user may wish to create a task with no explicit state objects, and `Run()` provides a way to do this.

5.4.5 Tasks with No State Object

There are many cases in which control tasks do not change state. For example, a DC motor might behave in the same way at all times, delivering torque in response to current no matter what. A simulated motor should reflect this behavior by never changing its behavior. A task which implements this simulated motor does not require any state structure.

In order to ease slightly the code overhead associated with writing one-state tasks, a method is provided to instantiate tasks without any states. To do so, the user overloads the `CTask` class's `Run()` method, replacing the default method's state transition code with the user's running code (such as a motor simulation). The result is a task which executes the user's `Run()` function whenever the `Schedule()` method determines that it's time for the task to be scanned. This technique allows smaller and simpler source files to be created for the same job.

A mechanism is also provided which allows the creation of *ad hoc* state structure in a task with no state objects. There is a member variable belonging to class `CTask` called `State`. This mechanism is provided for backward compatibility with earlier versions of the scheduler and is *not* recommended for use in new programs. Many useful debugging features in TranRun4 rely on state objects. If one is tempted to use the `State` variable to implement transition logic, it is a strong indication that it's time to create some state classes!

5.4.6 Creating Task Classes

Tasks are usually used as placeholders which serve to organize and schedule their states. Therefore, it is often not necessary to derive custom classes from class `CTask`. Instead, one simply instantiates objects of class `CTask` in the `UserMain()` function, attaches states to them, and it is ready to run.

In other cases, it may be useful to create custom task objects. For example, the PID controller class `C_SISO_PID_Controller`, in files `TR4_PID1.hpp` and `TR4_PID1.cpp`, is implemented without a state object, with all functionality in the derived task class. The `C_MotionP` class, in files `MotionP.hpp` and `MotionP.cpp` from the `Glue04` directory (see section 16.8), uses the task class to store data which is used by all the states within the task. Creating a task class is a matter of creating a descendent class to `CTask`:

```
class C_MotionP : public CTask
    {
    private:
        C_SISO_PID_Controller* ThePID;      // PID controller used by this profiler
        double Xcurrent;                    // Current position of the setpoint
```

```
        double Xtarget;                 // Position to which we're moving
        double StatusCode;              // Status code other tasks can see
        double CruiseSpeed;             // Maximum speed at which we 'cruise'
        double MaxAccel;                // Max. acceleration/deceleration level
        double CurrentAccel;            // Acceleration being used at the moment
        double Velocity;                // Current velocity of setpoint motion
        double Direction;               // Direction of travel, 1 or -1

        C_MotionP_Idle* IdleState;      // Here we declare pointers to all the
        C_MotionP_Hold* HoldState;      // state objects which belong to this
        C_MotionP_Accel* AccelState;    // task class
        C_MotionP_Decel* DecelState;

        double StopPoint (double, double,  // Function to find where a
                         double);           // hypothetical profile would stop
    public:
        // Constructor just calls the sample time task's constructor
        C_MotionP (CProcess*, char*, int, real_time, C_SISO_PID_Controller*,
                   double, double);
    };
```

The motion profiler task shown here holds pointers to its states as member data. This is not necessary in most cases; it is common for the state pointers to be created and kept in the `UserMain()` function instead. However, keeping state pointers as member data of a task class allows the states to be instantiated within the task's constructor instead of within the `UserMain()` function. This leads to slightly more modular code, which aids in the development of larger projects. The states are instantiated in the constructor as shown here:

```
C_MotionP::C_MotionP (CProcess* aProc, char*aName, int aPri, real_time aTime,
                     C_SISO_PID_Controller* aPID, double aSpeed, double aAccel)
    : CTask (aProc, aName, SAMPLE_TIME, aPri, aTime)
    {
    // Save the parameter
    ThePID = aPID;

    // Create the member states of this task and specify one as the initial state
    IdleState = new C_MotionP_Idle (this, "Idle");
    HoldState = new C_MotionP_Hold (this, "Holding");
    AccelState = new C_MotionP_Accel (this, "Accel");
    DecelState = new C_MotionP_Decel (this, "Decel");
    SetInitialState (IdleState);

    // Initialize the member data of the task
    CruiseSpeed = aSpeed;
    MaxAccel = aAccel;
    Xcurrent = 0.0;
    Xtarget = 0.0;
    StatusCode = 0.0;               // This means idle
    Velocity = 0.0;
    CurrentAccel = 1E-6;
    Direction = 0.0;                // Not travelling either way
    ...
    }
```

A disadvantage to instantiating the states inside the task constructor is that since the task keeps the state pointers as `private` data, member functions of the states cannot be called directly by outside tasks. In order to allow the states to be accessed, one would have to make the state pointers `public` and call states' member functions using an ungainly syntax:
`TaskPointer->StatePointer->StateMemberFunction()`.

Note that the version of the `CTask` constructor for sample time tasks was called. This signifies that this is an intermittent task.

5.4.7 State Objects

As we last left the long sequence of function calls which originated in the master scheduler's `Go()` function, the task's `Run()` function had called the current state's `Schedule()` function. The state's `Schedule()` function will finally call the `Entry()`, `Action()`, and `TransitionTest()` functions which contain the code written by the user. These are the functions which implement the transition logic structure which was described in section 3.7. A copy of the state class definition is shown next, with large sections removed for clarity.

```
class CState
    {
    private:
        int NonOverloadedCount;   // Count of functions not yet overloaded
        bool FirstTimeRun;        // Is this the first time this state's run?
        bool EnteringThisState;   // True when entering this state
        ...

    public:
        // Constructor is given a name for the new state and parent task pointer
        CState (CTask*, const char*);
        virtual ~CState (void);

        // Entry function is called when the parent task enters this state; Action
        // function runs repeatedly while remaining in this state; Transition Test
        // function is called at end of every scan and returns the value of a new
        // state to be entered next scan (or it returns NULL for no transition).
        // Schedule() is the function which calls these functions.
        virtual void Entry (void);
        virtual void Action (void);
        virtual CState* TransitionTest (void);
        CState* Schedule (void);

        void SetSampleTime (real_time);   // Function resets interval between runs
        real_time GetSampleTime (void);   // To get sample time of parent task
        ...
    };
```

The state's member variables are used to implement transition logic and ensure that the user has correctly supplied an entry, action, or transition test function to be run. If a state has been created for which the user has not supplied an `Entry()`, `Action()`, or `TransitionTest()` function, the scheduler will exit with an error.

Although one or two of these functions may be missing—for example, a state with no transition test function simply won't ever transition to another state—a state with none of these functions at all simply makes no sense.

The functions `SetSampleTime()` and `GetSampleTime()` are shown here as examples of a number of mirror functions which duplicate functions that belong to tasks. This is done purely for programming convenience. The scope of the task's member functions does not extend to its states (after all, a C++ compiler has no idea that tasks and states are related). But it is convenient to allow a task's member functions which control timing, data communication, and other utility functions to be used directly from within the state. So functions with the same names as those of the task's member functions are provided within `CState`. When these functions are called from within the state's scope, an internally held pointer to the parent task is used to call the task's member functions.

That part of state class's `Schedule()` function which runs in simulation mode is shown next. A large set of conditionally compiled code (implemented with `#ifdev` commands) is used to compile this function in different real-time modes, and the code for nonsimulation modes has been removed to make this listing somewhat readable.

```
CState* CState::Schedule (void)
    {
    CState* TheNextState;            // Points to state to which we transition

    // If this state is being entered, run the Entry() function
    if (EnteringThisState == true)
        {
        EnteringThisState = false;
        Entry ();
        }

    // We're remaining within this state, so run the Action() function
    Action ();

    // Run the transition test function and save the next state
    TheNextState = TransitionTest ();

    // If a transition has been called for, set the entering variable true so the
    // Entry() function runs next time this state's Schedule() function is called
    if (TheNextState != NULL)
        EnteringThisState = true;

    // Return a pointer to the state which will be run the next time around
    return (TheNextState);
    }
```

This is the code which implements the state scan shown in figure 3-6, with the small difference that exit functions are not defined in the TranRun4 structure. If the user wishes to create exit code associated with specific transitions, the exit code must be written into the `if/else` structure of the `TransitionTest()` function.

5.4.8 Creating State Classes

The implementation of a state is a fairly simple matter of deriving a class from class `CState`, adding any needed member data, writing a constructor, and overriding at least one of the entry, action, or transition test functions. An example of a state class is taken from the Glue project module `Glue01` in section 16.5. In the header file `GlueOven.hpp`, the state class is defined as follows:

```
class C_GlueOven_On : public CState
    {
    private:
        double AmbientTemp;          // Temperature in the room
        double OvenTemp;             // Oven temperature
        double ThermCoeff;           // Heat transfer coefficient "k"
        double Current;              // Electrical current in the heater coil
        double CoilResistance;       // Resistance of the heater coil
        double HeatCapacity;         // Heat capacity of the oven system
        double PreviousTime;         // Save time for Euler integration

    public:
        // The constructor sets default parameters and parent task of this state
        C_GlueOven_On (CTask*, char*, double, double, double, double);

        void Entry (void);           // We'll overload the Entry() and
        void Action (void);          // Action() functions of class CTask
    };
```

In this simulation, only one state is needed to model the oven. All the data defining the simulation belongs to the "oven on" state, and the state's parent task is instantiated as an object of class `CTask`—no custom task class need be created.

The state class's constructor saves data which was supplied as arguments to the constructor. This data—oven name, ambient temperature, and simulation parameters—is given as constructor arguments so that many oven objects, each with different parameters, can be created at once. Some code related to the communication of data to other tasks has been removed for clarity; intertask communication functions are discussed in chapter 6.

```
C_GlueOven_On::C_GlueOven_On (CTask* aTask, char* aName, double tAmb, double tCoef,
                              double coilR, double hCap)
    : CState (aTask, aName)
    {
    AmbientTemp = tAmb;              // Save the parameters into member data
    ThermCoeff = tCoef;
    CoilResistance = coilR;
    HeatCapacity = hCap;

    CreateDataInbox ("Current");     // For communication with other tasks
    CreateGlobalData ("Temp", &OvenTemp);
    CreateGlobalData ("Current", &Current);
    }
```

The entry and action functions have been overloaded for this state; the entry function is shown next. The `GetTimeNow()` function is used to retrieve a measurement of the current real or simulated time. This function is implemented within the scheduler's timer object and made globally available throughout the program. When a progam is written in a mode which uses simulated time the timer object will keep an artificial time counter, incrementing the time count periodically. When the program is changed from simulation mode to real-time control mode, an appropriate version of `GetTimeNow()` which measures time using clock hardware is substituted. This change is discussed in more detail in section 4.7.

```
void C_GlueOven_On::Entry (void)
    {
    PreviousTime = GetTimeNow ();      // Save previous time for action function

    OvenTemp = AmbientTemp;            // Begin with oven at room temperature

    CopyGlobalData ();                 // Send data to other tasks
    }
```

5.4.9 The Main File and UserMain() Function

Standard text-mode C and C++ programs have an entry point function called `main()`. This is the function which is first called when a program is run; when `main()` returns, the program's execution is terminated. This main routine is used for initialization and cleanup of the TranRun4 scheduler, and the user is not allowed to edit `main()`. Instead, a function called `UserMain()` is supplied which is the user's editable entry point to a scheduler program. The code inside this function follows a simple pattern, shown here in pseudocode:

```
Configure the master scheduler
Instantiate and configure process, task, and state objects
Call TheMaster->Go() to run the scheduler
Save data and clean up
```

In every scheduler project, there should be one set of main files. These files contain global data (in the header file) and the `UserMain()` function (in the source file). The main header file is used to hold all the global data items which are shared between source files. This should in general be limited to a set of task and state pointers. For convenience, the main header can also contain `#include` directives which cause it to include the header files which are associated with the various tasks in the project. A simple main header file, `GlueMain.hpp` from the Glue0 directory, is shown next.

```
#include "BeltAxis.hpp"
#include "GlueLogr.hpp"

//--------------------------------------------------------------------------------
// Global Pointers
//      These are pointers to the task and state objects used in the project.  They
//      are declared 'extern' here and declared and instantiated in GlueMain.cpp.
```

Section 5.4. Simulation in C++ (with TranRun4)

```
extern CTask* BeltAxis1;                            // Belt-axis task objects and
extern C_BeltAxis_Running* BeltAxis1_Running;       // states belonging to those tasks
extern CTask* BeltAxis2;
extern C_BeltAxis_Running* BeltAxis2_Running;
extern CTask* GlueDataLogger;                       // Data logging task object
extern C_GlueDataLogger_Running* GDLRun;            // and its states
extern C_GlueDataLogger_Stopped* GDLStop;
```

The **extern** declarations are used to make the task and state pointers available to all files (and hence all tasks) in the project. This allows any task to communicate with any other. All variables which are declared **extern** in the header file must be declared in a non-**extern** fashion within the source file. The source file GlueMain.cpp from Glue0 is shown here.

```
#include <stdio.h>
#include <tranrun4.hpp>
#include "GlueMain.hpp"

//-------------------------------------------------------------------------------
// Global Pointers
//      These are pointers to the task and state objects used in the project. They
//      are declared 'extern' in GlueMain.hpp and declared and instantiated here.
CTask* BeltAxis1;                           // Belt-axis task object
C_BeltAxis_Running* BeltAxis1_Running;      // State belonging to that task
CTask* BeltAxis2;
C_BeltAxis_Running* BeltAxis2_Running;
CTask* GlueDataLogger;                      // Data logging task object
C_GlueDataLogger_Running* GDLRun;           // and its states
C_GlueDataLogger_Stopped* GDLStop;

//===============================================================================
// Function: UserMain
//      This is the startup function called by the scheduler before scheduling begins.
//      It's used in a way similar to the main() function in regular C/C++ programs.
//      We use it to configure the scheduler timing, processes, tasks, and states.

int UserMain (int argc, char** argv)
    {
    // Set the master scheduler's timing parameters
    TheMaster->SetTickTime (0.001);
    TheMaster->SetStopTime (2.5);

    // Create a process in which the tasks will run. We use just one process for
    // this example, running in one environment on one computer
    CProcess* TheProcess = TheMaster->AddProcess ("The_Process");

    // Instantiate a belt axis task, specifying a priority and sample time
    BeltAxis1 = new CTask (TheProcess, "Belt Axis 1", SAMPLE_TIME, 10, 0.020);

    // Create a state for the task. Parameters: Task, Name, J, Kt, f, R, r
    BeltAxis1_Running = new C_BeltAxis_Running (BeltAxis1, "Axis1run", 1.0, 2.5,
                                    0.1, 10.5, 0.1);
    BeltAxis1->SetInitialState (BeltAxis1_Running);
```

```
    // We have to send a message to the class telling it to set the current,
    // so just ask the belt axis class to send the data package to itself
    BeltAxis1->SendDataPack (BeltAxis1, "Voltage", 0, 3.0,
                             MB_OVERWRITE, MB_NO_RECEIPT);

    // Repeat the belt axis creation process for another belt axis task
    BeltAxis2 = new CTask (TheProcess, "Belt Axis 2", SAMPLE_TIME, 10, 0.020);
    BeltAxis2_Running = new C_BeltAxis_Running (BeltAxis2, "Axis2run", 1.2, 2.3,
                                                0.15, 11.0, 0.1);
    BeltAxis2->SetInitialState (BeltAxis2_Running);
    BeltAxis2->SendDataPack (BeltAxis2, "Voltage", 0, 3.0,
                             MB_OVERWRITE, MB_NO_RECEIPT);

    // Instantiate a data logger task and its states
    GlueDataLogger = new CTask (TheProcess, "Data Logger", SAMPLE_TIME, 5, 0.050);
    GDLRun = new C_GlueDataLogger_Running (GlueDataLogger, "Running", "GlueData.txt");
    GDLStop = new C_GlueDataLogger_Stopped (GlueDataLogger, "Stopped");
    GlueDataLogger->SetInitialState (GDLRun);

    // Run the scheduler.  This makes all the real-time stuff go
    TheMaster->Go ();

    // Print some diagnostic information about what ran, when, and for how long
    TheProcess->DumpStatus ("ProcessStatus.txt");
    TheMaster->WriteProfiles ("TimeProfiles.txt");
    TheMaster->WriteTrace ("TL_Trace.txt");

    // A return value of 0 generally indicates everything worked OK
    return 0;
    }
```

The process object is instantiated first. Task objects are instantiated next, each task being given a pointer to its parent process. States are created immediately after the tasks and given pointers to their parents.

At some time before the master scheduler's Go() function is called, the tick time and stop time of the master are set. The tick time is used to control the rate at which simulated time is incremented; the timer will increment the time count by the tick time whenever a state function is called. When the simulated time reaches the stop time, the Go() function will exit and the program will run to completion. A function belonging to the master scheduler called Stop() is available to cause the scheduler to stop running at any time; this method can be called from anywhere within the program by invoking TheMaster->Stop(). It is common for a simulation to end on some event such as a process value being reached or an error condition occurring.

Note that although the **new** operator was used to create the task and state objects, there is no call to **delete** to destroy them. This is because the master scheduler automatically deletes all processes, tasks, and states which have been placed into the scheduling list. This feature is purely a matter of programming convenience. Any other objects which were created with **new** must be deleted, however, to prevent memory leaks which can cause a great deal of trouble.

5.5 Real-Time Realization with C++

Two timekeeping mechanisms are provided with the C++ programs:

Internal time (TIMEINTERNAL)

Free-running timer (TIMEFREE)

The selection of one of these modes is made in tasks.hpp; the relevant section is (from the pwm1 directory):

```
// Define the timing mode  -- One (and only one) of the
// following should be active
// #define TIMEFREE     // Use the PC's free-running timer
   #define TIMEINTERNAL // Use internal time calculation only
// (i.e., calibrated time). TICKRATE sets the internal
// time increment.

// This is the time granularity if the free running timer
// is not used. Each scheduler scan, the time is incremented
// by TICKRATE. You can adjust this value to make the time
// a decent approximation of the real time.
#define TICKRATE 0.0002
```

The internal time mode is used for simulation. It can be used to create a calibrated time mode by experimenting with the time interval (TICKRATE) until the actual run time matches real time. As noted previously, this is very easy to implement but not very accurate. It is useful for control systems that don't have very demanding timing requirements and/or are implemented in a computational environment which does not have easy access to a hardware timer.

The free-running time mode makes use of a timer that counts pulses from a hardware clock which automatically rolls over (resets itself to zero) when it reaches the end of its count. This timer is described in more detail in chapter 7. As long as the clock is sampled with a period less than the rollover period, the time duration from the previous sample can be computed. There is no need for the sampling to be done on a uniform basis; sampling can be done at any convenient time as long as the maximum period restriction is never violated. As implemented in the sample code, the sampling is restricted to being less than one-half the rollover period. With this restriction, sampling periods that are too long can be detected (on a statistical basis). The time resolution of the free-running timer is less than $1\mu s$, which is faster than the control software can actually make use of. This results in a full rollover time of 55 ms. There is a #define in clock.hpp (in the common directory of grp_cpp.zip) to pick which clock will be used.

Although not implemented here, it would also be possible to use the hardware timer that is supported by the operating system. This is accessed via calls to ftime(). However, its granularity is about 55ms on a PC which is running MS-DOS or Windows 95 or Windows 98, which is too crude for most control problems.

Chapter 6

INTERTASK COMMUNICATION

Communication among tasks is an inherent part of control software. Intertask communication is used both to make data generated in one task available to others and to synchronize the activities of different tasks. The design concerns in structuring the communication facilities include data integrity, communication efficiency, and software portability. These must be addressed for data that is explicitly defined by the user or programmer as well as data that is integral to the task model. The latter data includes task status information, state information, and all the other data which is used to provide task/state functionality.

It may be that a designer has chosen to meet the constraints of several high priority tasks through the use of several independent processors, one for each such task. This guarantees immediate attention to the highest priority task in each processor. Rather than requiring the resource shifting associated with real time scheduling, this technique adds additional resources to meet additional requirements. Naturally, the increased complexity brings additional problems: The tradeoff in using multiple processors is that data exchange must go through some external medium to get from one task to another. This has consequences of both overhead to service the medium, and delay for the actual transfer of information. These costs must be balanced against the savings due to less scheduling overhead and the potential increase in total processing resource.

A designer may choose to run all tasks in a single process but assign each task to a separate, asynchronously executing thread within that process. Since the tasks within different threads share the process's address space, normal language syntax such as function calls or external variables can be used for data transfer between tasks. However, if any form of preemptive scheduling is used there is a risk of data corruption. For example, if a task which is reading some data from a globally accessible variable in memory is preempted by a task which writes new data to that variable before the reading is complete, the data which is read may be corrupted. Methods must be specified to prevent such an occurrence. Different methods must be used depending on whether the tasks are running in different threads within the same process or within different processes.

Communication among tasks can be thought of as "push" or "pull" depending on whether a task actively sends information to another task ("pushing") or actively retrieves the information ("pulling"). The information flow direction is the same, but the task initiating the activity changes. Several communication mechanisms will be introduced which utilize "push" structures, "pull" structures, or a combination thereof. Three communication methods are described. The first, data transfer via function calls, is useful only for single-process jobs. It makes maximum use of the facilities built into programming languages while preserving data integrity. It is simple to use and doesn't require any specialized software. The other two methods, message passing and the distributed data base, are suitable for multiprocess applications. They make minimal use of language syntax but do require underlying specialized software to implement the data transfer.

6.1 Communication Within a Process

For applications designed *only* for single process implementation, communication is most easily handled using the language syntax. This preserves the maximum portability, particularly if language translation is necessary. Because communication only involves moving data within a common address space, efficiency is not a major issue. However, programming constructs that maintain data integrity are very important.

6.1.1 Data Integrity

Data integrity must be examined at three levels:

- Assuring that individual data values are correct

- Assuring that sets of data values have consistent context

- Assuring that data used within extended calculations are consistent

The first problem, that of the integrity of individual data values, arises in single process systems that use preemptive dispatching. Assume, for example, an assignment statement, x = y; in a low priority task. Assume also that the variable y can be set in a task of high priority that is dispatched preemptively. Interrupts preempt computational operations at the *machine instruction* level. That is, the interrupt allows the currently executing machine-language instruction to complete before taking effect. But the C language statement x = y; may require several machine language operations to complete. There is a chance that the high priority task will interrupt and change the value of y at the point where only *part* of y has been copied to x. After the high priority task finishes, the remainder of the copy operation will take place, but now using the new value of y. The result can be an x value that is corrupted and bears little resemblance to either the new or old values of y. For the sake of illustration, consider x and y to be 8-bit signed integers (2's-complement) in a computer that moves data 4 bits at a time. The initial value of y is 1, or in binary, 00000001. Suppose that just after the transfer of data begins, an interrupt occurs and the high priority task changes y to -1, which in binary is 11111111. The transfer is then completed after the high priority task has finished. There are two possible results, depending on whether the first four or last four bits are transferred first. If the first four bits (the high-order bits) are transferred first, the sequence of values taken on by x is:

$$XXXXXXXX \rightarrow 0000XXXX \rightarrow 00001111$$

Here we use Xs to represent the original bit values of x. If the last four bits (the low-order bits) change first, the sequence is:

$$XXXXXXXX \rightarrow XXXX0000 \rightarrow 11110001$$

If there were no interrupt, y would be 1 throughout the transfer, and the value would be correctly transferred from y to x. If the high-order bits were transferred first, the sequence would be

$$XXXXXXXX \rightarrow 0000XXXX \rightarrow 00000001$$

and if the low-order bits are transferred first, the sequence would be

$$XXXXXXXX \rightarrow XXXX0001 \rightarrow 00000001$$

In the preemptive case x ends up as either 15 (00001111) or -15 (11110001), depending on the order of transfer. In either case, the result is far removed from either the initial value of y or its changed value. If this error took place when our code was controlling a mechanical system, the result could easily be an actuator value that caused a violent motion of the controlled system, possibly damaging the machine or injuring someone.

The second data integrity issue, assuring the consistency of data sets, arises from a similar cause. In this case, data associated with an array, table, or data structure is communicated across tasks. Again, if an untimely interrupt occurs, some of the data in the set will be associated with an older version and some will be associated with the new version. In this case, although all of the data values may be correct, the set of data may not be.

For example, consider a control system whose gains are stored in an array. A lower priority task computes the gains and stores them in the array, and a higher priority task runs the control loop using these values. It may be the case that the arrays of gains which are created by the lower priority task allow stable control of the system, but the control gains which result from reading the array as it is being changed cause the controller to become unstable, causing damage or injury.

The third data integrity issue arises when a variable is used several times in a sequence of C (or other language) statements. For example, the following short sequence is commonly used to apply a limit to a variable:

```
x = some_calculation ();
if (x > max) x = max;
```

If this code were in a low priority task and a high priority task used the value of x, an interrupt occurring between these two lines of code would cause the high priority task to use the value of x computed before the limit was applied. This is an intermediate result — it was never intended that it be made available outside the function in which it resides. The preemption violates that intention and the high priority task gets an incorrect value with possibly serious consequences.

These potential problems will not always occur. It requires that the preemption happen at just the right (or wrong!) moment. The interrupt responsible for the preemption, however, is usually asynchronous with respect to the low priority code that is originally executing — in other words, we can neither predict nor control which lines of code are running when that interrupt occurs. The probability for an error to occur might actually be quite small. Thus, it may be unlikely that the error will be caught in normal lab testing, but very likely that it will show up in a production system.[1] This is an absolutely unacceptable situation. It can only be avoided by properly recognizing when such situations can occur and providing suitable protection against such an occurrence.

6.1.2 Design Rules

Two design rules that can help guard against data corruption problems are:

- Identify and protect critical regions

- Identify and isolate all data exchanges among tasks

Through the proper application of these rules, software can be written which is protected from data corruption by task preemption.

Critical Regions and Mutual Exclusion

Critical sections of code must execute atomically, that is, without preemption. This can be done by preventing preemption while the critical code is executing — by disabling interrupts or by disabling the dispatching of tasks that could engage in data exchange. The latter technique is used in both the group-priority and TranRun4 software packages. A paired set of function calls is provided that will prevent any user-defined tasks from preempting the running task. Interrupts which service system functions such as the measurement of real time are not prevented from executing. The functions used to define a critical section are `CriticalSectionBegin()` and `CriticalSectionEnd()`. These functions must be used with care. The primary rule for their usage is as follows: *Keep critical sections as short as possible.* In effect, what these functions do is to elevate the section of code they surround to the highest possible priority, thus preventing nominally higher priority tasks from

---

[1] It has been noticed in real-time programming classes that these errors are seldom seen during debugging sessions, but they almost *always* seem to occur when the "finished" product is presented for a grade. This effect is popularly known as the "demo demon."

running. This usage is a form of mutual exclusion because it prevents access to data (or other resources) by more than one task.

The use of a critical section can solve the first data-exchange problem—preventing data corruption. The assignment statement could be protected as follows:

```
CriticalSectionBegin ();
x = y;
CriticalSectionEnd ();
```

More care needs to be taken to prevent the corruption of sets of data, however. These broader data integrity issues can be taken care of by maintaining a separate set of variables that are used for exchange purposes. The point of data exchange can then be carefully controlled — by using only local variables for all calculations within the task the corruption of these data by other tasks can be prevented. For example, the following code is used in the group priority scheduler example program `heatsup.cpp` (in directory `heater1`):

```
...
setpt = temp_soak;
CriticalSectionBegin ();
xsetpt = setpt;
CriticalSectionEnd ();
...
```

The setpoint is copied from an internally computed variable and then is transferred to an exchange variable, `xsetpt`. (By convention, variables used for exchange purposes use `x` as the first letter of the name.) Exchange variables are never used except within a critical section, so the basic data integrity is always preserved. Note that the computation which computes the value of `setpt` is *not* placed within a critical section. All the variables with which this calculation works are local variables that may not be changed by any other task; and since this computation may take a long time to complete, we do not want to prevent higher priority tasks from executing during the calculation.

Data Exchange By Function Call

When data are being exchanged between tasks within the same process, it is often convenient to transfer the data by means of function calls. This permits the code responsible for data exchange to be isolated within one function. It also follows the principles of object-oriented design; all variables are encapsulated within the task objects, and data transfer takes place only through function calls which are specified as part of the interface of the task classes.

It should be noted (and will be repeated) that use of function calls or other data transfer means that rely on language facilities are dependent on a common address space for the tasks on both sides of the transfer. Although the method is simple and fast, it severely limits portability since tasks running in separate processes do not share a common address space.

It is possible to transfer data through the use of global variables implemented as `static` or `extern` variables. However, this technique allows the operations used to transfer data to be spread throughout the code, violating the rule of isolating data exchanges between tasks. Global variables also violate the principles of object-oriented programming by weakening the encapsulation of data within objects. Their use tends to make program maintenance and debugging more difficult and so the use of global variables for intertask communication is not recommended.

Data transfer functions can be used in either a push or a pull mode. Functions with names like `GetSomething()` implement the pull mode. The task needing the data calls the function at the point where the data is needed. Functions with names like `SetSomething()` implement the push mode. A task uses such a function to set a variable value in another task. In either case, the exchange variable is used inside these functions and in a critical section. The following example is also from `heatsup.cpp`:

```
void HeatSupTask::SetTempFinal (double new_tf)
    {
    CriticalSectionBegin ();
    xtemp_final = new_tf;
    CriticalSectionEnd ();
    }
```

This allows another task (in this case, the operator interface task) to set the value of final temperature that is used within the supervising task. Another example illustrates the get style function:

```
double HeatSupTask::GetTempFinal(void)
    {
    double rv;
    CriticalSectionBegin ();
    rv = xtemp_final;
    CriticalSectionEnd ();
    return (rv);
    }
```

In both cases, since only exchange variables are used, the tasks can control explicitly when the new information is actually utilized or when the internal information is made available for publication to other tasks. The final temperature information referred to in the previous example is used in state 3 of the supervisory task,

```
...
CriticalSectionBegin ();
temp_final = xtemp_final;
setpt = temp_final;
xsetpt = setpt;
CriticalSectionEnd ();
...
```

In this instance, the connection between the setpoint and the final temperature is also made within the critical region and the new setpoint value is made available to other tasks.

The get/set functions have a curious property with respect to task dispatching. Although they appear as member functions of the task whose data they access, they run as part of the task making the function call. Thus when the supervisory task in `heatsup.cpp` makes the function call

```
HeatCtrl->SetSetpoint(setpt);
```

the function is running as part of the supervisory task despite the fact that it is defined in the class associated with the control task.

6.2 Communication Across Processes

If an application is compiled and linked to run in an environment with independent processes (including the case of multiprocessors) then conventional programming methods for interchange of information among functions – through function arguments, return values, and global variables – will not work because the independent processes do not share memory address spaces.

Instead, a set of utilities must be provided which perform the operations necessary to transfer data between processes. Since a major goal of a unified real-time programming methodology is to retain code portability, a standardized user-level interface to the interprocess communication facilities must be supplied. The lower level utility software can then be customized for specific communication environments without requiring revision of the user-level code.

Common communication media include shared external memory (for multiple processors that share a bus, for example),[2] networks of all sorts, and point-to-point connections which transfer data from one processor to another. These media have different software interface protocols. If portability is to be preserved, the details of these protocols must be hidden from the user. Different communication mechanisms will have significantly different performance and cost, as well as different distance and environmental constraints, which must be factored into the final system design.

6.2.1 Message Passing

Message passing is one technique by which data may be transported between tasks in different processes. Here, we use message passing as a general term to describe the procedure in which a collection of information is transported in some sort of package from one task to another.[3] Although message passing is made necessary

---
[2] Shared external memory is not the same as the memory which is used within each process. As external memory, it cannot be accessed through the use of ordinary variables and function parameters; instead, it requires the use of special interprocess communication utilities.

[3] This encapsulation of data into *packages* resembles the encapsulation of data into network *packets*, but the concept is intended to be more general. Data packages may be transported by means of network packets, or by some entirely different means such as external shared memory or even the local memory used within one process.

by the need to pass information between tasks which reside in different processes, it can also be used to transport information between tasks in the same process. Although this use of message passing incurs some overhead, it may be convenient to use message passing to transfer data between tasks regardless of whether or not they reside in the same process. Then if some tasks are moved from one process to another, the code which transfers data between them will not have to be rewritten.

There is no general rule for what is in a package. The format could be unspecified so that the receiving task would have to decode it from a sequence of bytes; or one or more fixed data formats could be used. Whichever is the case, the information can be transformed into any necessary format as it moves from the sender to the receiver to allow passage through any communication medium that is being used. Message passing is, of necessity, a push type process. The task with the information must send a message to the task needing the information.

6.2.2 Message Passing in the Group Priority Scheduler

Samples and implementation details in this section are given only for single-process implementations. The code developed using the message passing and global database methods is completely portable to multiprocess environments. Multiprocess examples and implementation details are given in chapter 11.

The examples that follow are drawn from the group priority sample set in archive file `grp_cpp.zip`. They are based on the heater control problem and can be found in directories `heater1`, `heater2`, and `heater3`. The use of message passing in the TranRun4 scheduler is discussed in section 6.2.3.

Message Format

The samples used in this section come from the directory `heater2`. The program is functionally the same as the program in `heater1` but uses message passing for all intertask communication.

A simple message passing format is adopted here. A message has three components:

- The fact that it was received

- An integer

- A double-precision floating point number

This simple format simplifies conversion of the message at both ends and also simplifies the function calls used to send and receive messages. It provides enough generality to solve large numbers of real problems.

When messages are used for task synchronization, all that is needed is the fact of arrival. The integer is useful for information such as commands, while the double allows the passing of data.

Message Boxes

When a message is sent, the information passes through whatever communication medium connects the tasks and then is stored into a message box belonging to the receiving task. Message boxes are defined by the `MessageBox` class:

```
class MessageBox
   {
   public:
      int FromTask;
      int FromProcess;
      int MessageFlag;
      int MessageAvailable;
      double MessageValue;
      MessageBox (void) { MessageAvailable = 0; }
   };
```

This definition is found in `basetask.hpp`, in the `common` directory of the archive.

When the message arrives at the receiving process and the target message box is identified, the `MessageAvailable` variable is checked to see whether there is an old message still in the box. This would be indicated by a `TRUE` value for the `MessageAvailable` variable, which is cleared when the task reads the information from the message box. There are two delivery modes:

- Don't deliver messages if the box already has a message
- Deliver them anyway

The mode is specified on the sending side. `MessageFlag` and `MessageValue` are the integer and double values that constitute the message content.

The number of message boxes available is the same for all tasks and is specified by the preprocessor symbol:

```
#define NUM_MESSAGE_BOX 20    // Number of message boxes in each task
```

The most convenient way to specify what each message box will be used for is to `#define` a set of symbols. The task class name is put first to avoid overlapping names. The message boxes for the PWM task, for example, are:

```
#define PWM_PERIOD 0
#define PWM_DUTYCYCLE 1
```

These are defined in `tasks.hpp` so that the definitions are available to all tasks. The definitions must be available to all tasks since the task generating the information needs to know the identity of the message box to which to send the information.

Message box definitions for generic tasks present an interesting problem because both the parent and the inherited classes can use mailboxes. The PID control is such an application in this program. The generic part uses the upper end of the message box range, leaving the lower end free for the inherited class. These definitions are from the generic part (`pid_ctr2.hpp`):

```
// Define message boxes for data input to PID Control
// The derived task should not have any message boxes since it has
// no run function but, just in case, these are defined at the top
// of the message box set!
#define PID_START NUM_MESSAGE_BOX-1
#define PID_SETPOINT NUM_MESSAGE_BOX-2
#define PID_KP NUM_MESSAGE_BOX-3
#define PID_KI NUM_MESSAGE_BOX-4
#define PID_KD NUM_MESSAGE_BOX-5
#define PID_MINV NUM_MESSAGE_BOX-6
#define PID_MAXV NUM_MESSAGE_BOX-7
#define PID_NEWGAINS NUM_MESSAGE_BOX-8
```

and this is from the inherited part (in `tasks.hpp`):

```
// Message boxes (be careful that these don't overlap with the
// message boxes used by the generic portion of this class)
#define HEATCTRL_TEMPERATURE 0
```

Message Box Usage

Sending a message is done by a function call to `SendMessage()`. The function prototype is:

```
void SendMessage (BaseTask *ToTask, int BoxNo, int Flag = 1,
                  double Val = 0.0, int mode = 0, int ReturnReceipt = -1);
```

It is a member of the class `BaseTask`.

The first two arguments are required: a pointer to the task to which the message is to be sent and the message box number to receive the message. Only those two arguments are needed for a message for which arrival is the information to be transmitted. The remaining arguments are:

Flag	The integer value to be sent
Val	The double value
mode	Whether the message should overwrite a nonempty mailbox
ReturnReceipt	Whether or not a return receipt is desired (the value of this argument is the mailbox number in the sending task to which the receipt will be sent indicating whether or not the message was successfully delivered)

The mode values are defined in `basetask.hpp`,

```
#define MESSAGE_NORMAL 0
#define MESSAGE_OVERWRITE 1
```

Data sent in overwrite mode is used for cases where the receiving task should always have the most recent value, whether or not the previously sent value has been read. This example from the operator interface task (`heat_op.cpp`) is typical:

```
SendMessage (HeatSup, HEATSUP_TEMPFINAL, 1, temp_final, MESSAGE_OVERWRITE);
```

It sends a value of the final temperature (temp_final) to the supervisory task.
Receiving a message is done with GetMessage(). Its prototype is:

```
int GetMessage (int BoxNo, int *msg_flag = NULL, double *val = NULL,
                int *from_task = NULL, int *from_process = NULL);
```

It is also a member of the class BaseTask.

The only required argument is the number of the box from which a message is to be retrieved. If that argument alone is used, the return value will indicate:

-1 for an error

0 for no message in box

1 if a message was available

When a message is retrieved the box is emptied, i.e., the MessageAvailable variable is set to 0.

Note that this is true even if the minimal form of GetMessage() is used. A return value of 1 indicates that a message has arrived.

The from_task and from_process arguments allow the receiving task to find out which task sent the message. The task is identified by a code number rather than a pointer; an alternate version of SendMessage() is supplied so that the code number can be used to send a return message.

A typical example of receiving a message containing data is:

```
if (GetMessage (HEATSUP_TEMPFINAL, &msg_flag, &msg_val))
    {
    // Update if new value received
    temp_final = msg_val;
    ...
```

This is the receiving side of the sample of a message with data given earlier. It comes from heatsup.cpp.

The supervisory task also has an example of a message used for synchronization with no content other than the arrival of the message:

```
SendMessage (HeatCtrl, PID_START, 1, 0.0, MESSAGE_NORMAL, HEATSUP_PIDCONFIRM);
```

This message is sent to the PID controller to signal it to start control operations. The corresponding code from the control task is (pid_ctr2.cpp):

```
if (GetMessage (PID_START))        // Wait for "start" message
    {
    NextState = 2;
    ...
```

Note that no data is retrieved — just the fact of the message's arrival.

The sending message requested a return receipt by specifying the message box for it. This is used as a partial handshake so that the supervisory task does not proceed until it knows the message to start the controller has been delivered. The code for this is (in heatsup.cpp):

```
if (rv = GetMessage (HEATSUP_PIDCONFIRM, &msg_flag))
    {
    // Make sure message was delivered
    if ((msg_flag != 1) || (rv == -1))
        {
        StopControl ("Control start messsage not delivered\n");
        ...
```

The return value of `GetMessage()` is tested to see when a message has arrived and is then tested to make sure there was no error. The return receipt mechanism sets the integer value (`msg_flag`) to 1 if the message was successfully delivered. Nondelivery indicates a system failure, so control is stopped.

Waiting for the return receipt must be done as part of the test portion of the state. It may *not* be done with a `while` loop since that would cause blocking. When the target task is in another process, the time to deliver the message and its return could be quite long (depending on the communications medium used).

Single Process Implementation of Message Passing

The examples shown are from a program implemented as a single process. The message passing format is used to provide portability to a multiprocess environment. Multiprocess systems are discussed in chapter 11.

The implementation is relatively simple for messages that stay within the same process. The message passing functions must know for which process the message is intended, and what the present process is. The present process is identified in `tasks.hpp`:

```
const int ThisProcess = 0;      // Process number
```

and the process associated with the task to which the message is being sent is identified by a protected member variable of class `BaseTask` (in `basetask.hpp`):

```
int ProcessNo;     // Process that this task is located in
```

This variable defaults to 0 in the constructor. These variables will be used in the next chapter to establish a multiprocess structure. Each task is represented in processes other than its resident process by a skeleton instantiation that contains enough information for interprocess communication to work.

The two versions of `SendMessage()` call an internal function to actually process the message,

```
void BaseTask::SendMsg (BaseTask *ToTask, int BoxNo, int Flag,
                        double Val, int mode, int ReturnReceipt)
    {
    int ReturnValue;
    int to_process = ToTask->GetTaskProcess ();
```

Section 6.2. Communication Across Processes

```
        if (to_process == ThisProcess)
            {
            // Message is local -- deliver it immediately
            ReturnValue = ToTask->AcceptMessage (BoxNo, Flag, Val, mode,
                                                TaskCode, ThisProcess);
            if (ReturnReceipt != -1)
                {
                // Send a return receipt
                SendMessage (this, ReturnReceipt, ReturnValue);    // The value
                    // returned by Accept is sent as the flag to the return
                    // mailbox, which is specified by the value of ReturnReceipt
                }
            }
        else
            {
            StopControl ("Messages to other processes not implemented yet!\n");
            return;
            }
        }
```

This function mainly sorts out whether the message is intended for a task in this process or in another process. If it is in this process, the message can be delivered immediately by `AcceptMessage()`. If it isn't, an error is flagged in this version. It then tests to see if a return receipt is required and makes a call to `SendMessage()` to send the return receipt. The return receipt can be delivered at this point because there is no communication delay within a single process — `AcceptMessage()` is not blocking and returns immediately with information as to whether the message was deliverable.

`AcceptMessage()` delivers the message directly, first testing to see if the message box is already occupied.

```
int BaseTask::AcceptMessage (int BoxNo, int msg_flag, double msg_val, int mode,
                             int from_task, int from_process)
    {
    // Processes an incoming message
    // Returns 1 if message successfully delivered, 0 otherwise
    if ((BoxNo >= NUM_MESSAGE_BOX) || (BoxNo < 0)) return (0);
    if ((mode < 0) || (mode > 1))
        {
        StopControl ("Unknown message mode\n");
        return (0);
        }
    CriticalSectionBegin ();
    if (mode == 0)
        {
        // Check to see if message box is empty
        if (mbox[BoxNo].MessageAvailable != 0)
            {
            CriticalSectionEnd();
            return (0);            // Can't deliver
            }
        }
    // Copy message information
```

```
mbox[BoxNo].FromTask = from_task;
mbox[BoxNo].FromProcess = from_process;
mbox[BoxNo].MessageFlag = msg_flag;
mbox[BoxNo].MessageValue = msg_val;
mbox[BoxNo].MessageAvailable = 1;
CriticalSectionEnd ();
return (1);                       // Successful delivery
}
```

`GetMessage()` is simpler since it only has to query the local message box information. It never has to worry about where the message came from.

```
int BaseTask::GetMessage (int BoxNo, int *msg_flag, double *val,
                          int *from_task, int *from_process)
  {
  /* Returns:
    -1 for error
     0 for no message available
     1 for successful retrieval
  */
  if ((BoxNo >= NUM_MESSAGE_BOX) || (BoxNo < 0)) return (-1);   // Error
  CriticalSectionBegin ();
  if (mbox[BoxNo].MessageAvailable == 0)
    {
    CriticalSectionEnd ();
    return (0);                   // No message available
    }
  // Get information
  if (msg_flag != NULL) *msg_flag = mbox[BoxNo].MessageFlag;
  if (val != NULL) *val = mbox[BoxNo].MessageValue;
  if (from_task != NULL) *from_task = mbox[BoxNo].FromTask;
  if (from_process != NULL) *from_process = mbox[BoxNo].FromProcess;
  mbox[BoxNo].MessageAvailable = 0;     // Clear message box
  CriticalSectionEnd ();
  return (1);                           // Success
  }
```

6.2.3 Message Passing in the TranRun4 Scheduler

The message passing structure in TranRun4 is designed to permit the transfer of data between tasks regardless of the processes in which the tasks reside. As in the Group Priority scheduler, the underlying details of the communications are transparent to the user. If the user eschews function calls for data transfer between tasks and instead uses only message passing and the distributed data base (discussed in section 6.2.6), tasks can be moved between processes with no changes whatsoever in the intertask communication code.

Message Format

Messages in TranRun4 are referred to as data packs. These packages are sent by tasks and received by inboxes belonging to the receiving tasks. Once a data pack has been received by a task's inbox, the receiving task can read the data pack from

Section 6.2. Communication Across Processes

the inbox. The act of reading a data pack from the inbox causes the inbox to become empty. A data pack contains two data items:

- An integer

- A double-precision floating point number

The presence of a data pack in an inbox is, of course, also a type of information. The scheduling system adds some further information to a data pack automatically, including the serial numbers of the task and process from which the data originated.

Inboxes belong to tasks. Each task can own any number of inboxes. Each inbox is uniquely identified by a name; the name is a character string. This string need not obey the laws of variable names; spaces and punctuation are permitted (within reason). Inbox names are case sensitive. If a data pack is sent to a task which does not have an inbox whose name matches the inbox name on the data pack, the data is simply lost — no error checking is provided. If the user requires confirmation that a data pack has been successfully received, some form of *ad hoc* handshaking must be programmed at the user level.

The process of sending a data pack can be thought of as analogous to the mailing of paper letters via postal mail. The sender writes a letter. The letter is placed into an envelope and deposited in an outgoing mailbox. (Fortunately, the TranRun4 scheduler does not require the use of postage stamps.) The letter is picked up from the outgoing mailbox by the postal service and stamped automatically with a return address. Once the letter has been delivered to the recipient's mailbox, it is up to the recipient to retrieve the letter, open it, read it, and decide what the contents mean.

The limitations on information which can be transmitted using data packs are admittedly rather restrictive. They have been imposed in order to keep the user interface and underlying code as simple as possible. However, it has been found that most control problems can be solved through the use of the data pack method in conjunction with the TranRun4 distributed database. Occasionally, users employ clever tricks to circumvent the limitations of data packs — for example, encoding short character strings within double-precision numbers. However, these techniques are not necessarily portable, and type-safe programming is rendered impossible; the use of such techniques is discouraged.

The following functions are members of class `CTask`, found in file `TR4_Task.hpp`. They are used to create an inbox which can receive data packs from other tasks, to check whether an inbox contains incoming data, and to read the data from the inbox:

```
void CTask::CreateDataInbox (char*);
bool CTask::CheckForDataPack (char*);
bool CTask::ReadDataPack (char*, int*, double*, int*, int*);
bool CTask::SendDataPack (CTask*, char*, int, double, MB_WriteMode, MB_Receipt);
bool CTask::AcceptDataPack (char*, int, double, MB_WriteMode, MB_Receipt, int, int);
```

Often the user would like to invoke data pack handling functions from within the member functions of state classes. This can be done because corresponding data pack functions have been created as members of class CState. These functions belonging to states are implemented as wrapper functions which call the corresponding member functions of the state's parent task. These wrappers are provided merely for convenience.

`void CTask::CreateDataInbox (char* name)`
This function is used to construct an inbox. The function's parameter specifies the new inbox's name. The user need not worry about deleting the inbox; this is handled automatically when the task's destructor is executed.

`bool CTask::CheckForDataPack (char* name)`
This function checks if a data pack is available in the inbox with the given name. It returns `true` if a data pack is available and `false` if not. The contents of the inbox are not affected.

`bool CTask::ReadDataPack (char* name, int* idata, double* ddata,`
`                int* ftask, int* fproc)`
This function reads a data pack, if one is available, from the inbox with the given name. Parameters `idata` and `ddata` are pointers to variables of type `int` and `double` into which the data in the data pack will be placed if data is available. If the inbox is empty, the contents of `idata` and `ddata` will not be affected. The parameters `ftask` and `fproc` are pointers to integers which hold the serial numbers of the task and process from which the data pack was sent. If any of the four data pointers, `idata`, `ddata`, `ftask`, or `fproc`, is NULL, then the `ReadDataPack()` function will discard the data associated with that parameter rather than saving it to a local variable. `ReadDataPack()` returns `true` if data was available in the inbox and `false` if not.

`bool CTask::SendDataPack (CTask* pTask, char* name, int idata,`
`                double ddata, MB_WriteMode mode, MB_Receipt recpt)`
This function sends a data pack to another task which is pointed to by pTask. If the other task is running within another process, the TranRun4 scheduler will handle the details of sending the data pack through the appropriate communications medium. The integer and double data items are given in `idata` and `ddata`. The parameter `mode` controls what will happen if the inbox to which the data pack is sent already contains unread data: if `mode` is `MB_OVERWRITE`, the existing data will be overwritten by the new data; if it is `MB_NO_OVERWRITE`, the new data will be discarded and the old data left intact. The `recpt` parameter is currently unimplemented and should be set to `MB_NO_RECEIPT` in all cases.

```
bool CTask::AcceptDataPack (char* name, int idata, double ddata,
                 MB_WriteMode mode, MB_Receipt rc,
                 int ft, int fp)
```
This function is used internally within the scheduler to place data packs into inboxes. It should not be invoked by the user's programs.

6.2.4 Distributed Database

A distributed database is an alternate way of implementing interprocess communication. Certain information in each task is selected to be available for the database. That information is transmitted to all processes, so that any task in any process can access it. This is a pull process, as distinct from message passing which is a push process. When a distributed data base is being used, tasks can pull any information they want from it at any time. The distributed database is often considered more intuitive than message passing for shared information since shared information is used where it is needed. With message passing, in contrast, the information has to be sent to all tasks that might need it. This can lead to somewhat cluttered code if information needs to be sent to many tasks. Also, message passing requires that duplicate transmissions between two processes must be made when information is sent from a task in one process to more than one task in another process. On the other hand, because all information is distributed to all processes, need it or not, message passing probably makes more efficient use of communication resources most of the time.

It is possible to use both message passing and a distributed database in the same program. It is often found that one method or the other is more convenient for a given exchange of data. In particular, when writing "generic" classes which implement commonly used tools such as PID controllers and PWM signal generators, one finds that the interface to these objects is simplified when message passing is used to send data to the objects and distributed data is used to retrieve data from them. It is appropriate to think of the two methods of communication as complementary rather than competitive.

Data Ownership

An important concept in maintaining data integrity is ownership of data. In order to avoid problems arising when more than one task is trying to change a data value, each data value is owned by only one task, and the owning task is the only task allowed to change the variable's value. If this restriction were not imposed, unpredictable behavior would result as two or more tasks tried to change the value of a variable at the same time. The value in the variable would depend on which task wrote the data last.

6.2.5 Distributed Database in the Group Priority Scheduler

Database Structure

The database structure consists of an array of doubles. This allows enough flexibility to solve reasonably complicated problems, while keeping the communication software simple enough to be manageable. There are two pointers allocated, one in the protected part of BaseTask:

```
double *PrivateGlobalData;   // Global data array -- must be copied
                             // to public version for use by other tasks
```

and one in the private part:

```
double *GlobalData;          // Global information for this task -- accessible
                             // to all other tasks via call to GetGobalData()
```

The actual data allocation is done in the constructor for BaseTask. Initial set up of the arrays is normally done from the task constructor with a call to CreateGlobal(). This example is from sim.cpp:

```
CreateGlobal (2);            // Create a global data base
```

GlobalData is put in the private part of BaseTask so that it cannot be changed from task code, all of which is inherited from BaseTask so has access to BaseTask's protected part but not to its private part.

The PrivateGlobalData array is strictly for local use. Data is put into the array in the course of computation, but it is not available to other tasks until it is copied to the GlobalData array with a call to CopyGlobal().

Data can be retrieved from any task's global data array with a call to GetGlobalData(). Its arguments are the task pointer and the index of the data wanted.

Entering Data Into the Database

Each task maintains a local copy of the data base variables that it owns. Information can be written to the local copy at any time, since it is not accessible to any other tasks. This is done with an ordinary assignment statement,

```
PrivateGlobalData[HEATSUP_SETPOINT] = setpt;
```

This example is from heatsup.cpp. It takes the first step in making the data available by putting it into the local copy of the data array for variables owned by this task. No critical region protection is required for this operation because the transaction is strictly local.

When a consistent set of data has been entered into the local database, it is copied to the global version and released for access by other tasks by a call to CopyGlobal(). This function has no arguments.

Since the database is distributed, releasing it for access by public tasks does not mean that all tasks will have access to the data immediately. Tasks in the same process will have access as soon as CopyGlobal() completes. Tasks in other processes will only have access when the new information is communicated to their process. The `CopyGlobal()` operation does not necessarily even start communication. It just makes the data available. Communication can occur in any number of ways depending on the nature and bandwidth of the communication medium. Delays in access and nonuniform access to data in multiprocess systems is an important part of the performance assessment for them.

Retrieving Data from the Database

A function call to `GetGlobalData()` is used to retrieve data from other tasks. Its arguments are a pointer to the task from which data is desired, and the index of the specific value to be retrieved. The following example is from `heat_pid.cpp`:

```
temp = GetGlobalData (SimTask, SIM_TEMP);   // Get temperature
                                            // from the simulation task
```

Critical section protection is built into `GetGlobalData()`, so critical sections need not be explicit in the user-written code. The only exception is where several closely related values are retrieved and it is important that they be coherent in time. This example from `datalog.cpp` shows such a situation:

```
CriticalSectionBegin();
val = GetGlobalData(HeatCtrl,PID_PROCESS);
mc = GetGlobalData(HeatCtrl,PID_ACTUATION);
set = GetGlobalData(HeatCtrl,PID_SET);
CriticalSectionEnd();
```

Where data values are known to be contiguous in the data base array, a version of `GetGlobalData()` could be written that returns a set of values eliminating the need for an explicit critical section designation.

Single Process Implementation of the Global Database

The single process version of the database implementation is even simpler than the message passing functions. The initialization function primarily allocates memory:

```
int BaseTask::CreateGlobal (int nGlobal)
    {
    PrivateGlobalData = new double [nGlobal];
    GlobalData = new double [nGlobal];

    if ((GlobalData == NULL) || (PrivateGlobalData == NULL)) return (0);
    for (int i = 0; i < nGlobal; i++)
        {
        // Default values
        PrivateGlobalData[i] = GlobalData[i] = 0.0;
        }
    GlobalSize = nGlobal;
    return(1);   // Success
    }
```

Entering data into the database only requires ordinary assignment statements as noted earlier. The `CopyGlobal()` function just copies data from `PrivateGlobalData` to `GlobalData`:

```
int BaseTask::CopyGlobal (void)
    {
    if (NumberOfProcesses != 1)
        {
        cout << "Multiple processes not yet supported\n";
        exit (1);
        }
    CriticalSectionBegin ();
    for(int i = 0; i < GlobalSize; i++)
        {
        GlobalData[i] = PrivateGlobalData[i];
        }
    CriticalSectionEnd ();
    return (1);
    }
```

Because `GlobalData` is in the private section of `BaseTask`, only actual `BaseTask` member functions can access it — not member functions of inherited classes. This guarantees that no task can change data owned by another task. It is technically possible for one task to change the value of `PrivateGlobalData` owned by another task. However, it is hoped that this would not be an easy accident to make, and there is little reason to want to do it.

`GetGlobalData()` simply reads the data from the global data array of the task from which data has been requested:

```
double BaseTask::GetGlobalData(BaseTask *pb,int k)
    {
    CriticalSectionBegin ();
    double v = pb->GlobalData[k];
    CriticalSectionEnd ();
    return (v);
    }
```

For the multiprocess case, vestigial instantiations of tasks in other processes will be in every process. Thus, getting global data will be the same as it is for the single process case.

6.2.6 Distributed Database in the TranRun4 Scheduler

A global data item mimics the behavior of a carefully used static global variable in C++. The task which owns the data is the only one which can change it. When the data has been changed, the new value is automatically transmitted to all other tasks. Other tasks can then read the current value of the variable at any convenient time.

Global Data Format

The global database implementation in TranRun4 allows a task to broadcast data of type `double` to all other tasks. Other types of data are not supported in order to keep the programming interface as simple as possible. Code which protects data from corruption due to mutual exclusion issues is hidden inside the scheduler; the details of data transmission across various different media are taken care of automatically as well.

Unlike message boxes, global data items must be associated with pre-existing variables. One can think of the global database as being an automatically maintained message board which allows the world to see what the values of chosen variables are at any given time.

Each global data item is distinguished by the task which owns it and by its name. Global data names work in the same manner as data pack names: They are character strings which can contain any valid characters and must match exactly (including case, spaces, and punctuation) to specify the same item.

Implementation

A global data item is created by a call to the function `CreateGlobalData()`. This call is usually made within the constructor of a given task class. The global variable name and the local variable with which the global item is associated are specified as in the following example:

```
CreateGlobalData ("Door Position", &door_pos);
```

In this example, a variable called `door_pos` is being attached to a global variable whose name is given in the text string `"Door Position"`. Note that in order to read this data item, one would have to specify its name exactly, so the strings `"door position"` and `"DoorPosition"` would not work.

Somewhere in one of its running functions (`Run()`, `Entry()`, `Action()`, or `TransitionTest()`), the task which owns the `door_pos` variable will compute a new value. New values will probably also be computed for other global variables as well. After all the new values have been computed, a call to the function `CopyGlobalData()` is made. This function causes all of the global variables' values to be copied from their associated local variables and transmitted to all other tasks.

Data which has been transmitted to the global database can be read by the function `ReadGlobalData()`. It will be read into a local variable which belongs to the reading task. In this example,

```
ReadGlobalData (DoorControlTask, "Door Position", &p_of_door);
```

there must be a global data item called `"Door Position"` which is owned by task `DoorControlTask`. This function call reads the value of the global variable and stores it in the local variable `p_of_door`. The local variable could be member data of a task or state class; it could also be a variable which exists only within a function.

Note that the global database transmits data between tasks, not states. When a call to one of the global database functions is made from within a state function (such as a constructor, `Entry()`, `Action()`, or `TransitionTest()`), code belonging to that state's parent task will actually be called. Similarly, one cannot read global data belonging to a state — the task pointer must be used.

Chapter 7

TIMING TECHNIQUES ON PC COMPATIBLES

There are two groups of algorithms described here for measuring time on a PC. One is a calibrated approach which measures the execution speed of the program and calibrates this speed with a real-time clock. The other uses the PC's real-time clock to measure time changes, and keeps track of the overall time by summing these small time measurements. Interrupts are also used to help ensure that the range of the real-time clock is not exceeded.

7.1 Calibrated Time

At a fixed point in the execution of the program, a counter is incremented. This would most likely occur at the beginning or end of an execution scan. The total count is then multiplied by a calibration constant to yield time. This constant is determined by measuring the running time of the program and dividing by the counter value. The average scan time of the program can therefore be used as a measure of true real time. This is exactly the same procedure used to keep track of time in simulations.

There are several drawbacks to such a scheme. One is that from the point of view of program execution, time does not increase smoothly. Because the only timing guarantee is of average time, and because timing is linked directly to program execution, time may increase more rapidly or slowly during certain program execution states. It is also possible to have instances when successive measurements of time will yield identical results. This may restrict the use of numerical techniques which use a short time difference. Division by zero or other illegal instructions could ultimately result. On the other hand, this method is extremely simple to implement.

It is possible that the characteristic time(s) of the control problems involved are large enough that this method can be used successfully. The advantages to such a timing method are then in the simplicity. No special hardware is needed for a real-time clock. This reduces the cost of the system, possibly substantially. The software overhead of such a method is also extremely low. Some execution time is spent running the time incrementing code, but this is very small. Compared to interrupt-based timing or I/O reads and writes associated with more complicated schemes (described later), the time required for the calibrated time method appears even smaller. This method can produce acceptable time estimates for medium length time periods; for short time periods run-time inconsistencies make the time estimate inaccurate while for long time periods small errors in the actual average run-time can accumulate to substantial errors.

7.2 Free-Running Timer

A more exact method for measuring real time takes advantage of the PC's real-time clock. Counters, which are clocked by a known, fixed frequency are available on the PC's I/O bus. A free-running timer is one for which the counter is pulsed continuously and rolls over after reaching its maximum value (much the way an automobile odometer rolls over after 100,000 miles). The number output by the counter at any given time will be directly related to the real time. If the counter had an infinite number of bits, then this number could be read and converted to find the time.

$$t = \frac{C}{f_{clk}} \qquad (7.1)$$

where C is the number on the output of the counter and f_{clk} is the clock frequency. Of course the counter will have a finite number of bits, so it will overflow at regular intervals. The clock can still be used to measure relative time differences between two events by successive readings.

$$\Delta t = \frac{C_2 - C_1}{f_{clk}} \qquad (7.2)$$

In the case of an overflow (rollover) between readings, the relative time between the two can still be measured provided that the size of the counter (C_{max}) is known. The same equation applies, but with the stipulation that the maximum time difference being calculated is less than the rollover time of the clock.

$$\Delta t_{max} = \frac{C_{max}}{f_{clk}} \qquad (7.3)$$

This result relies on the nature of integer subtraction as it is implemented on almost all computers (2's complement arithmetic). The result can be verified by performing an integer subtraction of two unsigned numbers by hand or in a simple program. For a simple 4-bit case, try subtracting 14 from 2, or, in binary

```
0010 - 1110
```

The result is 0100, or 4, which is the number of clock ticks which would have occurred to count from 14 rolling over to 2. That is $15 - 0 - 1 - 2$. It is important to realize that this result will only occur if the number of bits in the counter is exactly the same as the number of bits of the numbers being subtracted. If the numbers 14 and 2 were converted to unsigned 16-bit integers and this subtraction calculated, the result would not be 4.

7.2.1 Hardware Timers on the PC

A standard peripheral on IBM PC compatible computers is an Intel 8254 programmable interval timer. This chip includes three 16-bit counters with programmable preset values along with some other mode options. The clock input to each of the timers is a fixed frequency signal of 1193180 Hz, or a period of approximately 0.838 microseconds.

DOS uses these timers during normal operation. Some uses of the timers are well standardized, while others are not. For instance, the use of Timer 0 appears to be standardized as the DOS time of day clock. Timer 0 is also unique in that it triggers an interrupt (IRQ0, interrupt number 07H) when it underflows. Neither of the other two timers (1 and 2) triggers an interrupt, so they are much more attractive for use as a free-running timer since there is little advantage to having an interrupt associated with a timer used this way. Unfortunately, testing has shown that the only timer which can reliably be assumed to operate in a standard way is Timer 0.

Timer 0 is used by DOS to keep track of the time of day while the computer is running. Normally DOS sets the preset value of Timer 0 at the maximum rollover rate (65535, 0xFFFF in hexadecimal, or about 54.9 milliseconds). This fact allows us to use Timer 0 simultaneously as a free-running timer.

Understanding how this timer works allows use of a simple yet very precise time measurement scheme. In this free-running timer mode, the value of Timer 0 is recorded whenever a program requests the current time. Equations 7.1, 7.2, and 7.3 can then be used to determine the difference in time since the last read of the timer and can increment the program time by that difference. As long as the timer is read once every counter period, or 54.9 milliseconds, then the absolute time can be determined with a resolution of 0.838 microseconds.

Although this scheme is very simple and effective, there are several things to watch out for. In particular, what happens if the timer is not read once every rollover period? This can be a very difficult thing to guarantee. Certain operations, especially on a slower system, could possibly take more time than this. Operating system calls, especially video calls or file I/O, can take tens of milliseconds to execute. Certain precautions must therefore be taken to avoid these situations.

Assume that for whatever reason there is a failure to read the clock within the allotted time. Since there is no other reference for the real time, there is no possible error check. A late read of the clock that should return a result slightly greater than 54.9 milliseconds will in fact return a much shorter time.

$$\Delta t_{returned} = \Delta t_{actual} - 54.9 ms \qquad (7.4)$$

The only way to try to detect such a timing problem requires defining that a time change of greater than some measurable value is an error. This means that instead of allowing the clock to be read once every 54.9 milliseconds, insist that it be read, for example, every 40 milliseconds. The 14.9 millisecond difference between the absolute maximum and artificial maximum serves as an error detection band. This at first seems to make the situation worse by requiring faster sampling of the clock, but an error checking scheme can now be implemented. If the clock is not sampled at or below the specified rate (40 milliseconds or less), there will sometimes be values of time difference measured that are between the false maximum of 40 milliseconds and the true maximum of 54.9. This will be defined as an error. This error will only be a statistical one, but it should eventually be detected. As a simple example of how this error detection works, consider a purely random clock sampling rate. The probability that a timing error would be detected for a given sample would be

$$P_{error} = \frac{T_{safe}}{T_{max}} = \frac{40}{54.9} = 0.728 \qquad (7.5)$$

Since the actual situation is likely not random, the time before an error is detected will vary.

The other two counters on the 8254 are identical in function to equation 7.5, so their use is also investigated. Timer 1 is designated for use by the BIOS to control memory refresh cycling. It would still be usable as a real-time clock as long as the period were known, but it could not be safely reprogrammed. Extensive testing has revealed that Timer 1 is being programmed by some unknown part of the operating system (BIOS or DOS) on some systems and that it cannot be reliably used as a real-time clock.

Timer 2 is designated as the PC speaker controller, so its use has also been investigated. Results with this timer are more encouraging than those for Timer 1. Apparently, the use of Timer 2 is not truly standardized for IBM PC compatibles. On some systems, Timer 2 works perfectly for measuring time in exactly the same manner as used for Timer 0. Unfortunately, some systems appear to reset Timer 2 as part of the keyboard handling routine, so real time can only be measured with this clock if no keys are pressed during the entire duration of the program. Considering that this is a nonstandard use of the timer, it is possible that other BIOS or DOS versions may exhibit similarly nonstandard behavior. Use of Timer 2 in the free-running mode is not recommended without thorough testing.

7.2.2 Performance Timers in Unix and Windows

Many modern operating systems contain code which implements an interface to free-running timers inside the host computer. Among these are Linux and Windows 95/98/NT/2000. These operating systems were not intended for real-time use—and neither were their timers, which were designed for use in measuring the execution times of regular application program code for such purposes as making word processors format text more quickly. Nevertheless, the performance timing functions inside these operating systems can provide a robust and reliable interface to the free-running timers in a PC.

The Windows 95/98/NT/2000 application program interface, or Win32 API, provides a pair of functions called `QueryPerformanceFrequency()` and `QueryPerformanceCounter()` which access the clock. The first returns the frequency of tick counts in Hertz; this is 1193180 Hz on a standard 80x86-class PC. The second returns a number of ticks which is guaranteed to increase uniformly as long as the operating system keeps running. The performance counter is used to compute intervals by simply subtracting a tick count taken at one time from a count taken at another. Since the tick count returned by `QueryPerformanceCounter()` is returned in a 64-bit integer, rollover is seldom a problem; the counter can count for approximately 490,000 years before overflowing.

It seems clear that these performance counters are merely an interface to the same hardware devices which have been used to measure time by the methods described in section 7.2. However, since the operating system assures that performance counters never overflow, the use of such an interface can save a great deal of time and trouble.

7.3 Interrupt-Based Timing

Although the free-running timer is an accurate method for measuring time, it is not robust to blocking code which prevents the clock from being read often enough. The timer interrupt on IBM compatible PC's can be used to increase the robustness of timekeeping in DOS applications. This method uses the same timer as the free-running timer method (Timer 0), but it reprograms the DOS clock interrupt and interrupt service routine (ISR) to help keep track of time.

The reason for using the clock interrupt to help with timing is that this interrupt will preempt many blocking procedures that would otherwise prevent the reading of the clock and proper time keeping. During the ISR, a call is made to the timer simply to read the current clock state. The same code which is used to calculate the time for the free-running timer can be used with this method, but the interrupt will make it almost certain that the clock will be read often enough (unless a higher priority interrupt is present which has a long execution time). Assuming that the interrupt occurs, time will be properly kept regardless of when or if the user program code reads the clock.

It is important to note that reprogramming of the DOS clock (Timer 0) will affect the ability of DOS to perform its normal functionality. In particular, DOS time of day services will not function properly while the ISR and frequency of associated with the clock are changed. It is possible to restore the ISR and frequency to the state present before they were reprogrammed, and this is recommended if the computer is expected to function normally after the user program terminates. However, the time of day will not be updated while the clock is being used by the user program, so it will certainly be incorrect even after it is reinstalled. This clock is also used to unload the heads on the floppy disk drive, so that function will also be affected if the floppy disk is used.[1] It is possible to keep these functions active by calling the DOS clock routine at appropriate times. However, for short prototype programs this is often not done.

The default clock interrupt is set to occur every full 16-bit clock rollover, or every 54.9 milliseconds. This value can be reprogrammed such that the timer is preset to any value less than 54.9 ms. The clock then counts down to zero, generating the interrupt on underflow. The clock immediately reloads the preset value and begins to count down again. When the preset value is changed, the interrupt occurs more often, decreasing the chances that the interrupt will be delayed long enough to cause an error. This approach is only advantageous if the free-running timer is calculated from a timer other than Timer 0 (Timer 2, for instance). In that case, the interrupt on Timer 0 is used to guarantee that Timer 2 is read frequently enough to avoid an error.

This approach will not work well if Timer 0 is used for both the interrupt and free-running timer. In order to cause the interrupt to occur at a time less than 54.9 milliseconds, the full range of counting on Timer 0 is no longer available. For example, if Timer 0 is set to generate an interrupt at half of the maximum period (27.4 milliseconds), then the free-running timer on Timer 0 must be read faster than 27.4 milliseconds. Decreasing the interrupt period has gained nothing in terms of robustness or accuracy compared to the longer interrupt period case. In fact, the free-running timer code must explicitly handle the fact that the rollover of Timer 0 in no longer 16 bits but is now 15 bits. Also, the efficiency of the code which runs an interrupt twice as often will be decreased because processor time is being wasted in the ISR which would otherwise be utilized by user code.

The best approach is clearly one which utilizes Timer 0 to generate interrupts at some fraction of the maximum clock frequency and Timer 2 as a free-running timer to actually measure time. The interrupt is then used to guarantee that the free-running clock is sampled frequently enough as to not cause a timing error. The simplest free-running time code can be used in this case because all 16 bits of timer data is used. The disadvantage of this method is that it is not portable to all PC's due to the nonstandard uses of Timer 2.

---

[1] Reprogramming the interrupt timer when running Windows is not recommended. The behavior of the Windows interface depends on this timer, and changing the timing causes the computer to behave in a less than usable (though somewhat amusing) manner.

Chapter 8

MULTITASKING: PERFORMANCE IN THE REAL WORLD

The universal real-time solution as outlined in section 3.9 depends on adequate computational speed to juggle all of the simultaneous activities without regard for the importance or computational characteristics of any of the individual tasks. The real world is not often so kind! Taking account of these real-world concerns, however, only requires modest additions to the design structure as it has been developed thus far. These additions allow for a stronger connection between the engineering system design and the computational technology so that system performance specifications can be met.

8.1 Priority-Based Scheduling—Resource Shifting

The essence of living within a finite computing budget is to make sure enough of the computational resource budget is expended on the highest priority tasks. The first addition to the design structure, therefore, is to associate priorities with the tasks. The job of the computational technology, then, is to shift resources away from the low-priority tasks and toward the high-priority tasks as needed to meet the given requirements.

How that is to be done depends on the criteria for success. There are at least two ways of measuring performance:

- Progress toward a computational goal

- Execution within a certain window

These two goals lead to different computational strategies and to fundamental task classifications.

8.1.1 Continuous vs. Intermittent Tasks

These two performance measures suggest a division of tasks according to performance type. Continuous tasks measure success by progress. They will absorb as much computing resource as they are given. Intermittent tasks respond to explicit events and have a limited computing function to perform for each event occurrence. Their success is measured by whether the response to the event occurs within a specified time of the event itself.

The universal solution outlined in the previous chapter uses only continuous tasks. The nature of a task's response to events is embedded in its transition structure, so no distinction needed to be made in task classification. To shift priorities most effectively, task dispatching can be associated with a variety of external events. The most common example is to associate dispatching with one or more clocks, but it can also be associated with external switch closures, keyboard or disk drive events, the arrival of information, and so on. This association only affects intermittent tasks; continuous tasks continue to absorb as much computing resource as they are given.

The computational complexity in a solution to a real-time problem depends strongly on how many different events are used to dispatch intermittent tasks and on how the computing resource is shared among the continuous tasks (the latter issue will be discussed later). Performance for intermittent tasks is normally specified by the delay between the occurrence of a particular event and the delivery of its result. Best performance will thus be achieved if such a task is dispatched by the occurrence of the specified event. However, dispatching all intermittent tasks this way incurs a substantial cost in computational complexity. The most cost-effective solution will reach some compromise that meets performance specifications with no more computational complexity than necessary.

Continuous tasks encompass that portion of a control job that is most like conventional computing in the sense that the goal is to deliver a result as soon as possible rather than at a specified time. They thus form the lower priority end of the real-time spectrum. Priority distinctions within the continuous tasks can be handled by, on average, giving more resource to the higher priority tasks. Within the universal approach, which is still applicable to the continuous tasks, priorities of continuous-type tasks can be recognized by giving high-priority tasks more scans than lower priority tasks.

A different approach is required for the intermittent tasks. In the simplest computing environment, that is, a single computing thread using a modification of the direct or universal approach described in previous chapters, intermittent tasks are distinguished from continuous tasks by one property: once they are triggered they are given as many scans as necessary to complete their activity.

This change alone is sometimes enough to meet performance specifications for a job that could not meet specifications using all continuous tasks. In particular, if the problem is that once triggered, the high-priority task does not get its output produced within the specified window, this will help meet specifications, because as an intermittent task it will be given all of the computing resource until it completes the response to the event. If, on the other hand, the problem is one of latency, that is, the task does not always start soon enough after the triggering event to meet the specification, then changes in the way tasks are dispatched will be required also.

8.1.2 Cooperative Multitasking Modes

Multitasking is a generic term in computing referring to the ability to do more than a single activity at a time. It is usually applied to single-processor configurations. In that context, since digital computers are by their nature only capable of doing one thing at a time, the simultaneity comes from rapid switching (with respect to the observer) from one activity to another, rather than truly simultaneous operation. In the context of real-time control software, this definition must be further refined to reflect both the engineering and the computational categories defined earlier.

First, to map from the engineering system description to the computational implementation, the engineering task is treated as an indivisible computing unit. Thus, each computing thread will contain at least one task.

Within any thread that contains more than one task, some form of dispatching must be implemented to share the thread's computational resource among its tasks. By default, the dispatching method used in the templates already introduced is to give equal treatment to all tasks, giving each task one transition logic scan at a time. This is also called a sequential dispatcher because it dispatches each task in strict sequence.

Since this dispatcher exists in one thread, it is referred to as a cooperative multitasking dispatcher. The term cooperative is used to describe the fact that it can never preempt a running program but can only perform its function when tasks voluntarily return control back to the dispatcher.

The success of cooperative dispatchers depends on the voluntary return of control, which in standard programming languages is difficult to design consistently. Cooperative dispatchers can be used very effectively in the transition logic context, however, because the state functions (entry, action, test, and if used, exit) are all specified as nonblocking. Thus, there is always return of control at the end of each state scan which doesn't require any special consideration on the programmer's part.

The next level of complexity in dispatching is to separate the continuous from the intermittent tasks and schedule them separately. The scheduler for the continuous tasks remains the same as it has been; the dispatcher for the intermittent task differs in that it gives each intermittent task as many scans as it needs before going on to the next task. This requires a change in the code within the intermittent tasks: each intermittent task must notify the dispatcher when it has completed processing the current event. It must also notify the dispatcher when it doesn't need any processing resource.

This is most easily done in the test function. Every test function in an intermittent task that is associated with waiting for the next event also notifies the dispatcher that it has no immediate need for CPU resource if the transition test result is negative. If the result is positive, no notification is given and the detected event will then be processed. It can also be done in other functions, if that is more convenient for the programmer.

In this breakdown into continuous and intermittent tasks the dispatcher does not have to know anything about the nature of the events that the intermittent events are processing. That is already taken care of in the state structure.

As noted, intermittent tasks often have a latency specification. Use of the sequential dispatcher can lead to longer latencies than can be tolerated because all other tasks must be scanned in succession before the dispatcher will return to a given task.

If the intermittent tasks as a group are assigned higher priority than the continuous tasks, a variant of that dispatcher can greatly reduce the latency seen by the intermittent tasks. In this dispatcher, all of the intermittent tasks are scanned after every scan of a continuous task. Thus, the maximum latency for the intermittent tasks (as a group) is reduced to the worst-case scan time for one continuous task scan.

This dispatcher, called a *minimum latency dispatcher*, can achieve the resource shifting needed to help a high-priority task meet its real time performance constraints; however, it takes significantly more overhead than the sequential dispatcher since many, if not most, of the scans of the intermittent tasks simply return the result that no computing resource is needed. The net effect is thus to improve the effectiveness of the high-priority tasks at the expense of the total amount of computing resource that is available for productive use.

If the intermittent dispatcher ends up running too often—that is, it produces a latency that is much lower than the maximum allowed—the overall system efficiency can be improved somewhat by establishing a minimum interval time for it. If less than the minimum interval has elapsed, the intermittent dispatcher is not run at all and the continuous tasks are serviced.

Selection of tasks to be included in the intermittent group is important in achieving system performance. The primary issue is latency. Any tasks that are not meeting their latency specification when scheduled in the continuous group should be included with the intermittent group. This is the high-priority group, however, so putting too many tasks in this group will make it difficult to meet the latency specifications. Tasks which are actually intermittent in nature but are able to meet specifications as part of the continuous group should normally be left there. Because this dispatcher does not know anything about the internal nature of the tasks, any task can be put in either group and will execute correctly. If a task is moved from the intermittent group to the continuous group, its completion notification will be ignored and the task will otherwise run normally.

8.2 Matlab Template for Minimum-Latency Dispatcher

The bulk of the code for the minimum latency dispatcher is unchanged from that of the sequential dispatcher version. An additional inner loop is added to the portion of the code that handles the tasks. The inner loop runs the intermittent tasks. It runs after every scan of the outer, continuous-task loop. The inner loop differs from the outer loop in that each task is run until it returns a TRUE value indicating its completion for this event.

```
for j = 1:ntasks_con
    tsk = tasks(j,:);    % name of j-th task
    feval(tsk);   % No arguments - task takes care of state
                  % and when it runs internally.
    inc_time;     % Increment time
    del = get_time - tbegin;  % Time since last simulation
    if del >= del_tsim       % If true, do a simulation step
        tbegin = get_time;
        [tt,xx] = odeeul('proc_sys',del,x,ode_opts);
        x = xx;   % Update the state vector
    end
    % Run the intermittent tasks - they all run for each scan
    for k = 1:ntasks_int
        tsk = tasks_int(k,:);    % name of k-th task
        flag = 0;    % Completion flag
        while ~flag
            flag = feval(tsk);
                  % Loop until task signals completion
            inc_time;   % Increment time
            del = get_time - tbegin;
                  % Time since last simulation
            if del >= del_tsim
                % If true, do a simulation step
                tbegin = get_time;
                [tt,xx] = ...
                    odeeul('proc_sys',del,x,ode_opts);
                x = xx;   % Update the state vector
            end
        end  % of while ~flag
    end  % k-loop
end  % j-loop
```

The only change in the code for the task is to add a return value for completion.

8.2.1 Example: Simulation of PWM-Actuated Heater

This Matlab simulation is an illustration where one task has much tighter latency requirements than the rest of the tasks.[1] In this case, the task that is responsible for PWM generation must run much more often than any of the other tasks.

The system simulated is a simple example of a thermal processing system that might be found in a variety of industrial settings. The control objective is to heat

---
[1] archive: grp_mtlb.zip, directory: heatctrl

an object to a specified temperature, hold it there for a specified amount of time, then bring it back to its original temperature.

The controller has three tasks:

- `heat_sup`: supervisor to generate setpoints

- `heat_pid`: feedback control of temperature

- `heat_pwm`: generate the PWM actuation

The system response is shown in figure 8-1. The top graph shows the PID controller output, which gives the duty cycle to the PWM generator. The bottom graph shows the temperature of the object (solid line) and the setpoint (dashed line). Because the PWM generator is given the same priority as other tasks, it runs slowly. The ripple from the relatively slow PWM (the PWM period is 1 time unit) shows up very clearly in the temperature plot.

When the period of the PWM is reduced there is less interaction between the PWM and the heater. As shown in figure 8-2, the ripple in the temperature due to the PWM is much reduced.

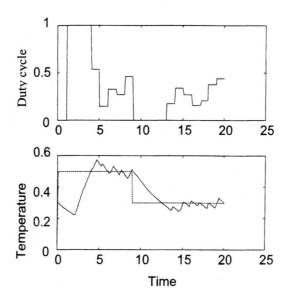

Figure 8-1. Heater system response, PWM period = 1.0

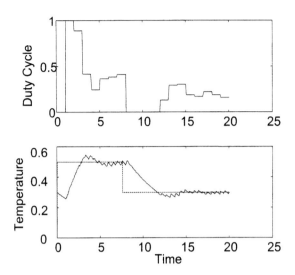

Figure 8-2. Heater system response, PWM period = 0.5

8.3 Cooperative Multitasking Using C++

The C++ version of minimum latency dispatching will be explored using the same heater control example that was used for the Matlab section.[2] Before introducing that example, however, the PWM example from the previous chapter will be revisited since it forms a major part of the heater control problem.

Minimum latency dispatching will be shown for the PWM problem, but it only has two tasks. In this case task latency will not be affected, however, using minimum latency dispatching will assure that the PWM task does not have to wait for the supervisor task to run before completing its operation. This happens at the beginning of every cycle, where the best PWM performance is achieved if the transition from Compute_Times to PWM_On takes place immediately. Variants of the PWM example will then be introduced to demonstrate how the use of the C++ property of inheritance can add code reusability to control system programs in a way that is easy to document and maintain.

The major change in implementing minimum latency dispatching is not in the program itself, but in the TaskDispatcher() function in basetask.cpp (in the common directory). In the previous chapter, the section of TaskDispatcher() dealing with sequential dispatching was shown. Here is the whole execution loop:

---

[2] archive: grp_cpp.zip, directories: pwm1, pwm2, and pwm3

```
        for (int j=0; j < jmax; j++)
            {
            for (int i=0; i < List2Num; i++)
                {
                if ((TempTask = List2->NextItem()) == NULL)
                        {// Get the next Task on the list
                        cout << "\nTried to access empty task list\n";
                        return(-1);
                        }
                Suspend = 0;
                while (Suspend != 1)
                        {
                        Suspend = TempTask->Run();
                        IncrementTime();
                        if (Suspend == -1)
                            {
                            cout << "\nProblem running Task: " <<
                                        TempTask->Name << "\n";
                            return(-1);
                            }
                        }
                }
            if(List1Num == 0)break; // No items in List1
            if ((TempTask = List1->NextItem()) == NULL)
                    {// Get the next Task on the list
                    cout << "\nTried to access empty task list\n";
                    return(-1);
                    }
            Suspend = TempTask->Run(); // Run the Task
            IncrementTime();
            if (Suspend == -1)
                    {
                    cout << "\nProblem running Task: " <<
                            TempTask->Name << "\n";
                    return(-1);
                    }
            }
...
```

List1 is the low-priority list of continuous tasks, treated by sequential dispatching; List2 is the high-priority list of tasks treated as intermittent tasks through minimum latency dispatching. As can be seen from the loop structure, each time one task from List1 is scanned, all of the tasks from List2 are scanned. The call to TaskDispatcher() is complete when all of the low-priority tasks have had one scan.

Each of the task lists is now maintained separately; the following code instantiates two lists, one for low-priority (continuous) tasks and one for higher priority (intermittent) tasks:

```
// Declaration of TList pointers are in tasks.hpp
LowPriority = new TList("Low Priority");
Intermittent = new TList("Intermittent"); // No min-latency
```

Section 8.3. Cooperative Multitasking Using C++

Tasks can be added to these lists with the **Append** method that is a member function of the `TList` class:

```
LowPriority->Append(PWM);
LowPriority->Append(PWMSup);
```

This is how the list is constructed in the example of the previous chapter (some intermediate lines of code have been removed for clarity). To change this so that the PWM task is treated as intermittent, these would become:

```
Intermittent->Append(PWM);
LowPriority->Append(PWMSup);
```

The only change in the task code is the addition of the flag indicating when the intermittent task completes its work. The following shows a short fragment including that computation(`pwm1.cpp`):

```
...
case 1: // State 1: "Compute_Times"
// Entry Section
if (RunEntry)
{
// only run this on entry to the state
// entry code goes here
// limit the value of DutyCycle to [0,1]
if(DutyCycle < 0.0)
DutyCycle = 0.0;
else if(DutyCycle > 1.0)
DutyCycle = 1.0;
EndOnTime = DutyCycle*Period+Time;
EndOffTime = Period+Time;
}
// Action Section
// no action code for this state
// Test/Exit section
// transition tests go here.
if(DutyCycle > 0.0)
NextState = 2;
// exit code goes here
// No exit section for this transition
else if (DutyCycle <= 0.0)
NextState = 3;
// exit code goes here
// No exit section for this transition
return (0); // run another state before yielding
// to other tasks
...
```

State 1 (`Compute_Times`) returns 0 indicating that other states should be run before yielding control. The other states all return 1 indicating that the task is done until the time for a transition occurs.

Running this in simulation mode (`TIMEINTERNAL` defined, `TICKRATE` defined as 0.01) gives the following fragment of the transition audit trail:

```
PWM, 2, 3, 1.600000
PWM, 3, 1, 1.640000
PWM, 1, 2, 1.650000
PWM, 2, 3, 1.690000
PWM, 3, 1, 1.770000
PWM, 1, 2, 1.780000
PWM, 2, 3, 1.800000
PWM, 3, 1, 1.900000
PWM, 1, 2, 1.910000
PWM, 2, 3, 1.950000
PWM, 3, 1, 2.030000
PWM, 1, 2, 2.040000
PWM, 2, 3, 2.120000
PWM, 3, 1, 2.160000
PWM, 1, 2, 2.170000
PWM, 2, 3, 2.290000
PWM, 3, 1, 2.310000
PWM, 1, 2, 2.320000
PWM, 2, 3, 2.440000
PWM, 3, 1, 2.460000
PWM, 1, 2, 2.470000
PWM, 2, 3, 2.550000
PWM, 3, 1, 2.590000
```

This shows the 0.01 second time granularity and also shows the minimum latency dispatching. Note that when state 1 executes, state 2 follows 0.01 sec later, the minimum possible.

When using this in real time with calibrated time mode, the step size (TICKRATE) is about 0.00004 sec (0.04 ms) to match real time (on a 50MHz 486). A segment of the audit trail shows the operation in this mode:

```
PWM, 2, 3, 2.623960
PWM, 3, 1, 2.703320
PWM, 1, 2, 2.703360
PWM, 2, 3, 2.705840
PWM, 3, 1, 2.803440
PWM, 1, 2, 2.803480
PWM, 2, 3, 2.806040
PWM, 3, 1, 2.903560
PWM, 1, 2, 2.903600
PWM, 2, 3, 2.924320
PWM, 3, 1, 3.003680
PWM, 1, 2, 3.003720
PWM, 2, 3, 3.053800
```

This shows similar characteristics but at a much smaller time granularity.

To run in real-time mode using the hardware timer, TIMEFREE must be defined in the tasks.hpp header file. This gives the following audit trail fragment:

```
PWM, 2, 3, 2.627563
PWM, 3, 1, 2.706867
PWM, 1, 2, 2.706974
PWM, 2, 3, 2.709557
```

```
PWM, 3, 1, 2.807078
PWM, 1, 2, 2.807185
PWM, 2, 3, 2.809764
PWM, 3, 1, 2.907306
PWM, 1, 2, 2.907412
PWM, 2, 3, 2.928219
PWM, 3, 1, 3.007567
PWM, 1, 2, 3.007674
PWM, 2, 3, 3.057839
PWM, 3, 1, 3.107836
PWM, 1, 2, 3.107944
```

This is now using actual measured time. The time to run state 1 can be seen here to be about 0.11 ms.

8.3.1 Inheriting Task Behavior—Two PWMs

Using the inheritance properties of C++ is very powerful for situations in which there is a need to create more than one object of the same type.[3] The PWM programs are a good example of that. If there is more than one device to be driven by PWM there should be a separate PWM task for each of these. If there is no difference between the tasks except data, all that has to be done is to instantiate as many copies as needed of the task. For example, several copies of PWMTask could be instantiated and all would execute independently. In the case of the PWM, however, this may not work because the PWM input and output often need unique functions to get at the appropriate I/O for each task. This requires one additional level of inheritance to create a class that has all of the properties of the class defined in the previous example but can also define its own unique versions of the PWM output functions. Both versions of this are shown here. The directory pwm2 contains example code for a program that generates two PWM signals by instantiating two tasks from the PWMTask class; pwm3 shows the use of inheritance to accomplish the same end for the case where independent PWMOn() and PWMOff() functions are needed.

There are two changes needed to create the program with two PWM tasks that differ only in their data. The first is to actually instantiate both tasks:

```
PWM1 = new PWMTask("PWM1");
PWM2 = new PWMTask("PWM2");
PWMSup = new PWMSupTask("PWMSup",0.4,0.2);
```

The second is seen in the instantiation of PWMSup—its constructor takes two numeric arguments instead of one. They specify the periods of the commands to the two PWMs.

Figure 8-3 shows a typical response. In this case, two different periods are used for the duty cycle commands. The actual PWM period is the same for both (although there is no need for that to be the case in general).

---
[3]archive: grp_cpp.zip, directories: pwm2 and pwm3

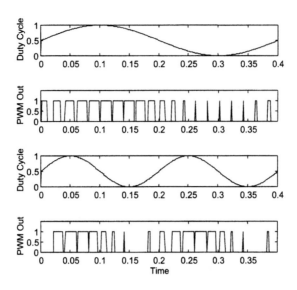

Figure 8-3. Two PWMs (by instantiating two tasks from the same class)

8.4 Preemptive Multitasking Modes

If the latency specifications still cannot be met, a preemptive solution will be needed. Such a situation could arise, for example, if the action functions of one or more low-priority tasks required substantial amounts of computing time. This does not violate the nonblocking condition, which states only that computing time in any of the state functions must be *predictable*, not that it must be short. In other cases, it might be necessary to violate the nonblocking condition. Though certainly not desirable, violations could arise due to the need to use software not under the control of the real-time programmer. Examples of these might be operator interface software or mathematical software with nondeterministic iterations.

Preemptive dispatching makes use of the computer's interrupt facility, which can, in response to an external event, temporarily halt the execution of whatever code is currently running and start some other code running. Since the interrupt facility is part of the computer's hardware, this preemption takes place on a much faster time scale than can be achieved by the minimum latency, cooperative dispatcher and so can meet much stricter latency requirements.

Interrupts are associated with a variety of events. In the case of mechatronic control systems, by far the most common interrupt events are from timers. Timers work by counting the pulses coming from an oscillator with a very stable frequency. The timer can be connected to the interrupt system by having it generate a logic signal whenever the counter counts down to zero. The range of the period depends

Section 8.4. Preemptive Multitasking Modes

on the frequency of the oscillator and the number bits in the counter register. A common arrangement is to have an oscillator frequency of about 1 megahertz and a counter with 16 bits. This gives a minimum counting period of 1 microsecond (too short to be used for interrupts) and a maximum of about 60 milliseconds. Depending on the importance (priority) and latency specifications of the tasks that respond to timer events, there can be any number of independent timer/interrupt combinations.

The other common interrupt source in mechatronic systems is from events in the target system. These events can be caused by switch closures, optical beam breaking, process variables exceeding a preset limit, etc. Again, there can be any number of independent interrupts of this type. Most of the other interrupt sources are associated with the computer system activities such as the keyboard, disk, network interface, etc., rather than control.

Preemption is used to establish a new computing thread. When an interrupt occurs, the current thread is suspended and a new thread is started. These threads share a common memory space but are otherwise independent entities. At the simplest level, the interrupt mechanism itself is a dispatcher, albeit a hardware-based dispatcher. Each interrupt-dispatched thread could contain only one task. This would be very desirable for tasks with very short latency tolerances. However, few computer systems have interrupt systems complex enough to devote a separate interrupt to each high priority thread, so interrupt-dispatched threads often contain more than one task. The tasks inside the interrupt-dispatched thread must be dispatched by a cooperative dispatcher.

Interrupt-dispatched threads must contain only intermittent tasks. Because the hardware interrupt establishes a priority level ahead of any software-based priorities, if the tasks in an interrupt thread were to ask for "continuous" CPU resource, all other tasks would be locked out. Thus, all of the continuous tasks must be left in the lowest priority (usually non-interrupt) domain; that is, they get to run whenever no interrupts are active. Threads that are dispatched directly by interrupts must normally be quite short. Because the interrupt mechanism on most computers is constructed in such a way that all other interrupts of the same or lower priority are locked out until the present interrupt is finished, it becomes essential to keep this time as short as possible.

With interrupts there are two major priority levels: those tasks running at hardware-interrupt priority and those tasks that run when no interrupt threads are active. There is an intermediate priority level, however. This consists of tasks that are dispatched as the result of an interrupt, but which run at a priority below that of any of other interrupt tasks. Such tasks can be preempted by interrupt activity, but on the other hand, no continuous tasks can run until these intermediate priority-level tasks all complete. As with the direct-interrupt tasks, tasks at this priority level must also be of intermittent type. Dispatchers based on interrupts come in two flavors:

- Those that operate through the function-call mechanism
- Those that manipulate the execution context directly

A cooperative dispatcher operating within an interrupt falls into the first category. It operates by calling the task function, which returns at the end of a scan. The case of an interrupt thread that has only a single task is really no different. In both cases, when the intermittent task (or tasks) finishes the interrupt ends. The intermediate priority level works by a similar mechanism, so it is also in the class of dispatcher that uses the function-call mechanism.

Both interrupts and function calls use a similar mechanism whereby suspended entities return to active status in the reverse of the order in which they were suspended. For functions, the function call does this. In a sequence of function calls the returns proceed from the most recently called function back to the originating functions. Likewise, with interrupts, the preempted activities resulting from a series of interrupts return to active status in reverse order. This works because the amount of information needed to restart a suspended activity (the computing context) is minimum when this order is followed.

Dispatchers that manipulate the full execution context are not constrained to follow the reverse-resumption order. Because the full context is known, tasks can be suspended and resumed in any order. For transition logic-based systems, this is most important in the way that continuous tasks are dealt with. In cooperative dispatching, the continuous tasks are given access to CPU resource according to the number of transition logic scans they use. This is true whether a particular task uses a large or small amount of CPU time for its scan. If this is an important consideration or if it becomes necessary to include blocking code as part of a system with several continuous tasks, the more complex context-based dispatching can be used. This permits round-robin dispatching, in which each continuous task is given a certain amount of CPU time and then control is passed to the next continuous task. This is also called time-sliced dispatching. In this case, the suspension/resumption order follows a circular list (circular because after the last task on the list is given its slice of CPU time, the first task then gets its turn) so the simpler dispatching method will not work. There are other properties that context-based dispatchers have, but these are less important than time-slicing in the transition logic environment. Simpler dispatchers based on function return mechanisms will be used throughout most of this text.

8.5 Realization of Interrupt-Based Dispatching

The cooperative dispatchers discussed in the previous section form the basis for interrupt-based dispatching. The steps are:

- Decide how many priority levels are necessary

- Decide which interrupt sources will be used

- Write a dispatching function for each interrupt source

- Attach each dispatching function to its interrupt

The goal of the realization method is to be able to change these decisions without having to rewrite any of the control (task/state) code. The only parts that have to be changed are the scan control loops. By maintaining portability in this manner, the same control code can be used for several phases of the development project, and implementation decisions can be delayed until more information has been obtained about system function and final performance specifications.

8.5.1 How Many Priority Levels Are Necessary?

Hardware priority levels allow for the satisfaction of latency requirements. Simple resource shifting can be handled by giving higher priority tasks more transition logic scans per pass than are given to lower priority tasks. If all latency specifications can be met with a cooperative dispatcher (with or without minimum latency dispatching) then only one priority level is necessary. No interrupts are required. Use of interrupts will only increase the cost and complexity of the solution and make the program *much* harder to debug and maintain.

Adding the interrupt priority level is usually the next step if the latency requirements can't be met. Tasks with very short latency needs and also short execution time for each event will work at this level. The execution times have to be short enough so that these tasks do not interfere with one another. This is necessary because interrupt-based tasks operate at the CPU's highest possible priority level and will lock out other tasks until they complete their activity.

If some of the tasks with short latency requirements also use more computing time than is consistent with running at interrupt-priority level, then the intermediate-priority level will be needed.

For the methodology used at this point prioritization will be limited to grouping tasks by priority level. No further priority assignments beyond that will be necessary.

8.5.2 Which Interrupt Sources Will Be Used?

This is a decision that ultimately can have significant economic consequences on the production system. Use of more interrupt sources than are needed adds to both the hardware and the software cost.

It is important to emphasize that there is never any intrinsic need for interrupts in a control system. Interrupts are used solely to improve performance.

With this in mind, the first thing to look at is the timer interrupt since that is the most commonly needed interrupt. If any critical tasks are time-based at least one timer interrupt will be needed. How many independent timers are needed must be determined from testing and measuring the worst case latencies for the critical tasks. Even if there are several time-based tasks, adequate performance may be obtained with just one timer.

The number of timers is a particular problem at the prototyping stage of a project. Standard desktop and laboratory computers are normally used for control of prototype systems in order to speed up the development process. However, these computers may be quite limited in the number of interrupt sources available. IBM-PC class computers, for example, typically have only one available timer interrupt. On the other hand, the general purpose computers used for prototyping tend to have faster processors than the target production computer which mitigates the latency problems.

The next most commonly needed interrupts are from logic signals. However, even if it appears that logic (digital I/O) interrupts are necessary, it may be possible to use the timer interrupts to cover the tasks based on digital I/O interrupts. This is because there will be many fewer tasks using the timer interrupt than were using the cooperative dispatcher, so it may be possible to meet the latency requirements without using the digital interrupt.

Another important note: tasks use internal logic in their transition tests to determine when and what actions to take. The event used to invoke the dispatcher for the task is irrelevant as long as the task can meet its performance specifications.

Other interrupts such as serial or network ports may be part of the control activity in the sense that information derived from that source is used by the control. Those interrupts are normally considered to be part of the underlying operating system, rather than the control system software. When working on a bare target computer, it may be necessary to write code for such devices, but even in that case, the basic computing functions should be separated from the control system software.

8.5.3 Interrupt-Based Dispatching Functions

A new general model for constructing the control program is needed when interrupts are introduced. In the templates described earlier, the main function (or a function derived from it) had the responsibility for initializing all of the control variables and task states, setting the output array variables, associating tasks with dispatching lists, and running the major execution-dispatching loop.

With interrupts present, this format must be modified. The main function does not change much, except that its execution loop only dispatches the tasks that are not interrupt-based. This includes all of the continuous-type tasks, and those of the intermittent tasks that can meet their latency specifications with a cooperative dispatcher.

Each interrupt source then has an associated ISR that is called via the interrupt mechanism. Other than taking care of the housekeeping associated with the interrupt, all it does is call `TaskDispatcher()` to execute its list of tasks. The timer ISR for use with PCs is located in the file `basetask.cpp` in the common directory.

One additional part of the interrupt control must be accomplished by dispatchers for the intermediate priority level. That is, before actually running its tasks, the dispatcher must reset the priority level of the hardware interrupt controller so that the high-priority interrupts will be allowed to occur even if an intermediate priority level task is running. The intermediate priority tasks are thus preemptible by the high-priority tasks. This requires both hardware and software to permit reentrant interrupts. A reentrant interrupt is one which occurs while code invoked by a previous interrupt of the same type is still running. For example, if there is a timer interrupt which results in an intermediate priority task to be started, another timer interrupt could occur before the intermediate priority task is done. Not all systems allow this. If a system's hardware or software does not support reentrant interrupts, intermediate-level priority cannot be used.

8.5.4 Attaching Dispatching Functions to Interrupts

Another new job for the main routine, as part of its initialization, is to connect each of the interrupt dispatching functions with its associated interrupt. While the details on how to do this differ from system to system, the general outline is the same.

The computer's interrupt system has a list of memory addresses, each of which tells the interrupt system where to execute code associated with that interrupt. This is called an interrupt vector table. When the interrupt occurs, the code that gets executed must first save all information needed to restore the computer to its state at the time of the interrupt, so the previously executing code can resume. Most commonly, some of that code is built in to the underlying operating system, and some of it must be written explicitly. Once the system state is thus saved, the interrupt dispatching function can be run. When it is done, the procedure is reversed and the computer is restored to its preinterrupt state. The previously executing code then resumes.

Chapter 9

A CHARACTER-BASED OPERATOR INTERFACE

Most mechatronic systems have an operator interface of some sort. In some cases it is as simple as an on/off switch; in other cases it is a complex combination of computer monitors and fixed switches, knobs, meters, and other elements. In all cases, a well-designed operator interface can improve the effectiveness, reliability, and safety of the system while a poorly designed one will have the opposite effect. The advent of computer-based control systems has opened new directions in operator interface design but at the same time has introduced new ways to confuse the operator. From a design point of view, building an operator interface from software has probably increased the proportion of the total design cost that goes to the operator interface. But from an economic viewpoint, the use of computer-based operator interfaces has probably significantly reduced installation costs because there are less components that need to be mounted and wired in a computer-based interface than in a conventional interface.

The topic of operator interface design is vast. It encompasses not only engineering and software design but also a great deal of psychology. The goal in this chapter, however, is quite limited: we develop an operator interface technology that is as portable as the rest of the control program and is suitable for use in prototyping. The reason for these very modest goals is that there does not appear to be a technology available that will meet our portability requirements and is simple enough to be feasible for use in prototyping.

The form of the operator interface used here follows the operator interface presented in Auslander and Tham[15], redone for C++ implementation. It is a somewhat simplified version in that the use of keyboard function keys is eliminated and control keys are substituted. This is because several environments that are important for prototyping do not support function keys.

9.1 Operator Interface Requirements

A mechatronic system operator needs unambiguous access to relevant information about the system and consistent command input to affect its behavior. While mechatronics systems can still make use of traditional, dedicated-function components for all or part of its operator interface, the discussion here will focus on that part of the interface implemented using computer display and input technology. For the record, dedicated operator interface components include switches, lights, alphanumeric displays, knobs, thumbwheels, and so forth. Since these are all dedicated devices, to the mechatronics control program they are handled in exactly the same way as other physical devices.

A standard computer (i.e., desktop) user interface has a two-dimensional display, a keyboard, and a mouse or other pointing device (and sometimes sound). The display has adapted well to the industrial world, but keyboards and common pointing devices are less desirable in the factory floor environment. They can be sensitive to dirt and other environmental problems, and they are not always easy to use to operate a machine. A common substitute is to overlay the display with a touch-sensitive device and use that for operator input.

9.2 Context Sensitive Interfaces

The most remarkable aspect of the switch to computer-based operator interfaces is the change from the traditional control panel, in which all information outputs and command inputs are available all the time, to computer-based interfaces in which an ever-changing subset of the interface is all that is available at a given moment.

Although it is theoretically possible to reproduce the complete-availability model in computer-based systems by providing enough displays and input devices to simultaneously show everything, this is rarely seriously considered. Instead, it was recognized that one or a small number of displays induced a very valuable point of concentration. If the proper information and command set could be provided, the operator could act much more quickly and accurately than with the very diffuse traditional control panel. This is a context-sensitive operator interface.

Use of context-sensitive interfaces brings both opportunities and risks. The opportunities discussed earlier, faster and more accurate operator actions, are very valuable. Another opportunity is that the complexity of the system being controlled is not limited by size or expense of the control panel since only the immediately relevant aspects of its operation need to be considered at any given moment.

The risks are significant also. The demands on the operator interface designer are much more stringent because each context must contain what the operator needs to effectively control that context. If it doesn't, there is little or nothing the operator can do about it! In addition, the consistency requirements of any type of operator interface must be considered. For knobs of the same type, for example, it is important that increase-decrease be associated with the same action of all of the knobs. That is, if turning the knob clockwise decreases the associated action in one of the knobs it would be confusing to the operator to find that the convention is reversed on another. Some analogous examples for everyday life are hot-cold water faucets, radio-TV volume controls, and light switches. Likewise, a computer display should use color, blinking, character size, and graphic elements in consistent ways.

9.3 User Interface Programming Paradigms

There are two major paradigms for user interaction with a computer. The older of these is the command line interface. In this method the computer presents the user with a prompt, a signal that the computer is ready to accept input. The user can insert any valid command at that point and the computer will do something. The best known interface of this type is the PC/MS-DOS operating system. The second method presents the user with a variety of choices. Picking any one of them will cause an action of some sort. The most familiar of these are the windowing-type of operating systems, Windows, Mac, Motif, and so on. Both of these types of user interfaces are in common use today and are found as the basic operator interface and within applications, often mixed. For example, under the PC/MS-DOS operating system many applications such as word processors present the user with a window-like interface. On the other side, for example, mathematical programs such as Matlab present the user with a command line interface while running under windowing operating systems.

There are also other crossovers between these paradigms that are relevant when considering the design of a mechatronic system operator interface. Design of a context-sensitive operator interface would seem to lean in the direction of the windowing-interface because it presents the operator with available choices. The command line method requires that the operator memorize the list of possible commands, a major training and reliability problem. On the other hand, standard windowing interfaces hide lots of information in covered up windows, branching menu structures, and so forth.

9.4 Mechatronics System Operator Interface

Some simple design rules will be propounded here for the design of operator interfaces for mechatronic systems:

1. All of the information and actions associated with a given context must be directly available to the operator.

2. Information and actions should be self-explanatory.

3. Operator interface context changes should take place only when requested by the operator. Emergencies represent a possible exception to this rule.

4. Operator interface code should follow the same nonblocking rules as other code.

5. The operator interface code should be portable across operating environments.

Unfortunately, no candidate fitting even this simple set of rules has been found. The problem is that operator interface construction software tends not to be portable and operates in environments that do not support real time software with the performance requirements typical of mechatronic systems.

The software selected for use here meets the portability and nonblocking rules, but provides only alphanumerical information and derives its actions only from control keys. It thus does not do a good job of making information and actions self-explanatory, as is often done using animated graphical displays.

The conclusion is that within the design philosophy presented in this text, operator interface implementation technology is an area requiring considerably more work.

9.5 Operator Interface Programming

The operator interface defines a screen or window which is displayed to the operator and contains information and commands for the present context. In an operating environment it is expected that only this screen will be visible to the operator. In the prototyping environment, there may be other items on the actual display screen at the same time.

9.5.1 The Operator Screen

There are three types of items defined:

- Input items, where the operator can enter information.

- Output items that display current data and status.

- Actions triggered by control keys.

As many screens as desired can be defined; each screen represents an operator interface context and the screens can be changed very easily from inside the program. The input and output items can be any of the common C or C++ data types: integer, floating-point, character, and so forth.

The geometry of the screen is fixed: the input items are on the left half of the screen, the output items are on the right half, and the control key items run across the bottom of the screen.

A test screen is shown in figure 9-1 (compressed to fit in a small area). It shows all of the relevant screen elements as they are viewed at the start of the program. This test screen comes from the `time_op` program described later in this chapter.

The arrow symbol "->" shows the point of data insertion. It is moved down through the list of input items by use of the Tab key, wrapping from bottom to top as it is moved past the lowest item. The arrow can be moved up with the key just above the Tab (usually the key with the tilde "~" as its uppercase character). When numbers are entered, the characters typed are entered to the right of the arrow. Either moving the cursor or clicking the Enter key causes the characters to be converted to a value and stored in the relevant variable.

Output items are updated automatically. No special user action is required.

Pressing an indicated action key (Ctrl+X or Ctrl+R in this example) causes the related function to be executed.

```
    Another Test Window

       An Integer:    ->    0         A Counter      8
       A Long Int:          0         Time           0.008
       A Float:             0
       A Double:            0
       Long Double:         0
       A String:            Blah


       Ctrl-X      Ctrl-R
       Stop        Reset
       Program     Counter
```

Figure 9-1. **time_op** **program test screen**

Programming for this operator interface is similarly simple. Input and/or output items are set up by specifying the address of a variable that is associated with the item, and a label to go with the item. Specification of an action key requires the choice of the control key (some combinations are not valid since they are already used for other operations), the label, and the name of the associated function. An initial call sets up the internal data handling. Refreshing the screen and handling keyboard input are done with nonblocking calls to an update function.

9.5.2 Programming Conventions in C++

The first example comes from the timer test program, which has a simple operator interface done for the purpose of testing its timing characteristics.[1] Its simplicity makes it a good example to illustrate setting up and running an operator interface. Following this example, the operator interface for the heater control example will be examined.

The initial step is to define a screen. This is done with a statement of the form (from `time_op.cpp`),

```
COperatorWindow* TheOpWin = new COperatorWindow ("Another Test Window");
```

The pointer, `TheOpWin`, must persist over the lifetime for which the screen it defines will be used. In the case of this example program it is an automatic variable in the main function. In most control programs, it will be a static variable in a task file.

The operator interface object having been created, all of the screen items associated with this screen can be defined and added to the screen:

---

[1] archive: `grp_cpp.zip`, directory: `time_op`

```
TheOpWin->AddInputItem ("An Integer:", &AnInt);
TheOpWin->AddInputItem ("A Long Int:", &A_Long);
TheOpWin->AddInputItem ("A Float:", &A_Float);
TheOpWin->AddInputItem ("A Double:", &A_Double);
TheOpWin->AddInputItem ("Long Double:", &A_Long_Double);
TheOpWin->AddInputItem ("A String:", A_String);
TheOpWin->AddOutputItem ("A Counter", &Counter);
TheOpWin->AddOutputItem ("Time", &Time);
TheOpWin->AddKey ('X', "Stop", "Program", DumpOut);
TheOpWin->AddKey ('R', "Reset", "Counter", ResetCounter);
```

Note that it is not necessary to specify the data types of the variables used. This is because the definitions of the various member functions have been overloaded to account for all supported data types.

The arrangement of the display on the screen depends on the order in which the items are defined: top to bottom for the input and output items, left to right for the control key items.

Initial display and activation of a window take place with a call to Display():

```
TheOpWin->Display ();
```

The input and output items are not displayed on the screen during this function call; subsequent calls to Update() actually paint the items on the screen. Display() causes the screen's title and key function information to be displayed, as these items do not need to be modified by Update() calls.

The display functions are nonblocking. Repeated calls to Update() are used to keep the screen refreshed and respond to operator actions:

```
while ((Time < EndTime) && (DoExitNow == 0))
    {
    TheOpWin->Update ();
    Counter++;
    ...
    }
```

Each call to Update() updates one of the input or output items and checks for any user input. If there is user input an appropriate action is taken. In a control program the initial call to Display() would normally be in an entry function and the call to Update() would normally be in an action function.

A screen is deactivated with a call to Close() when the real-time program is finished or when another operator interface is about to be drawn in place of the first.

```
TheOpWin->Close ();
```

There are two action keys in this example, one to reset the counter and the other to cause an exit from the program. Pressing an action key causes the specified user-defined function to be called. In this case, the functions are:

Section 9.5. Operator Interface Programming

```
static int DoExitNow = 0;

void DumpOut (void)
   {
   DoExitNow = 1;
   }
static int Counter = 0;

void ResetCounter (void)
   {
   Counter = 0;
   }
```

These functions just set values for variables. Other action key functions can do more complex operations if desired, although the need to keep their execution time short should be observed.

There are two sources of latency in the calls to `Update()`: the time it takes to update an output item plus do minimum input processing and the time it takes to run an action key function. The processing time is a function of the run-time environment and is not within the direct control of the programmer. The action functions, however, are in control of the programmer and so should be made as short as possible.

A minimal version of `tasks.hpp` is included with the program to provide all of the appropriate definitions and included files.

9.5.3 Heater Control Operator Interface

The following example creates an interface to control a heater program using the group-priority scheduler.[2] A similar example using the TranRun4 scheduler is described in section 16.7.

The heater control program has a two-screen operator interface. The first is activated before the real-time operation starts and is used to enter initial operating parameters. The second is active while control is going on and is handled by a separate task devoted to operator interface. The generalization of this is that any program can have multiple screens for the operator interface. Additional states in the operator interface task take care of starting and closing the relevant screens. There is no limit to the number of screens that can be used as long as the use of the screens is clear to the operator. It is very easy to get overly enthusiastic about multiple screens and have a structure that is too complicated to understand readily during system operating conditions.

The pre-real-time operator interface is done with a function that is included in the same file as the operator interface task code (`heat_op.cpp`). It is configured as an ordinary function that communicates with the heater control class through its public functions. It should probably have been done as a member function, although since it can set parameters in a number of tasks, it would have to use the public interface in any case. That function and its associated control-key function are:

---

[2]archive: `grp_cpp.zip`, directory: `heater1`

```
static int setup_done;

void SetupSetDone (void)
    {
    setup_done = 1;
    }

void HeatSetup (void)
    {
    double kp, ki, kd, mn, mx, tickrate;
    COperatorWindow* OpWinSetup = new COperatorWindow ("Setup Initial Data");
    HeatCtrl->GetGains (&kp, &ki, &kd, &mn, &mx);

    // Get current values
    tickrate = GetTickRate ();
    OpWinSetup->AddInputItem ("Tick Rate", &tickrate);
    OpWinSetup->AddInputItem ("Kp", &kp);
    OpWinSetup->AddInputItem ("Ki", &ki);
    OpWinSetup->AddInputItem ("Kd", &kd);
    OpWinSetup->AddKey ('X', "Start", "Control", SetupSetDone);
    setup_done = 0;
    OpWinSetup->Display ();                      // Initial display
    while (!setup_done) OpWinSetup->Update ();
    SetTickRate (tickrate);
    HeatCtrl->SetGains (kp, ki, kd, mn, mx);
    OpWinSetup->Close ();                        // Shut down the window
    }
```

The function `HeatSetup()` is called from the `main()` function prior to starting the real time operation. The structure of this function is typical of all operator interface functions – it first gets the current values of relevant values, then sets up the screen for display, then sets the new values. The only difference between this version and the versions that run while control is active is that the latter need to consider updating to new values while the screen is still active rather than waiting until the screen has been closed as in this case.

The screen that is used during control is set up in the constructor for the task class:

```
// Constructor for the class
OperatorInterface::OperatorInterface (char *aName)
    :BaseTask (aName)
    {
    // Create screens
    OpWin1 = new COperatorWindow ("Heater Control Operation");
    OpWin1->AddInputItem ("Final Setpoint", &temp_final);
    OpWin1->AddOutputItem ("Time", &Time);
    OpWin1->AddOutputItem ("Setpoint", &setpoint);
    OpWin1->AddOutputItem ("Temperature", &temperature);
    OpWin1->AddOutputItem ("M-Cntrl", &mcntrl);
    OpWin1->AddKey ('X', "Stop", "Program", Stop);
    DeltaTaskTime = 0.05;                        // Set refresh interval
    NextTaskTime = 0.0;
    State = 0;
    temp_final = 0;                              // Make sure it has a legal value
    }
```

Section 9.5. Operator Interface Programming

The variable `OpWin1` is a member variable of the class so it will persist as long as the instantiation of the class persists. This is important since it must be used in the task's `Run()` function as well.

The screen display is controlled in the `Run()` function. In this case there is only one state, since there is only one screen to control:

```
...
switch (State)
    {
    case 1:                             // Display main operation screen
        // Data from the controller
        HeatCtrl->GetData (&temperature, &mcntrl, &setpoint);

        // Entry Section (empty)

        // Action Section
        Time = GetTimeNow ();
        temp_final = HeatSup->GetTempFinal ();
        tf_old = temp_final;
        OpWin1->Update ();

        // Check for user input and refresh outputs
        if (tf_old != temp_final) HeatSup->SetTempFinal (temp_final);

        // Transmit the new value

        // Test/Exit section

        // No other states yet!
        break;

    default:                            // check for an illegal state.
        cout << "\nIllegal state assignment, Operator Int. Task\n";

        return (-1);
    }
...
```

There are a couple of operating characteristics which should be noted. The variables that are changed must be updated while the screen is still active. However, the screen is scanned many times per second whereas the user will only rarely actually change the value of the variable. If the update is done anyway, that will generate a lot of thrashing as variables that haven't changed are updated. To avoid this waste, a local variable is kept which can be tested to see if there has been a change. The thrashing problem is particularly acute if the new variable must be transmitted over a network or some other remote mechanism.

The second issue is how often the screen should be updated. There are two issues here—CPU usage and screen readability. Screen updates are fairly expensive in terms of CPU usage, so reducing the frequency is useful. It also turns out that reducing the frequency is useful for readability since a very rapidly changing screen is hard to read. An update rate of 10 to 20Hz seems to work reasonably well (keeping in mind that each scan updates only one screen item, so a complete screen update takes as many scans as there are screen items).

Because the screen setup is done in the constructor and the variable pointing to the screen instantiation is a member variable, in a multiscreen application this same screen can be opened or closed as many times as needed without going through the setup step again.

Chapter 10

GRAPHICAL OPERATOR INTERFACES

The operator interface presented in the previous chapter provides the necessary functionality for early prototype use and has the important properties of ease of use and portability. However, even a prototype intended for demonstration outside of the immediate development lab needs to have an operator interface that takes advantage of graphical display capabilities.

The problems traditionally associated with graphically based operator interfaces are:

- Lack of portability

- Very difficult to program

- Computationally very intensive

The operator interface technology presented in this chapter, while not necessarily solving all of these problems, at least side-steps them sufficiently to be useful in many situations.

The lack of portability and the intense computational demands are met in the same way: using a separate processor to handle the operator interface. The enabling technology here is the local area network (LAN). LANs are now well standardized and interfacing from almost any computer to a LAN is straightforward. LANs cause delays in intertask information transmission, but the operator interface does not generally place response time demands on the system that are beyond the capability of readily available LAN technologies. The network technology discussed in the chapters on distributed computing can be used to realize the LAN portion discussed here by running the operator interface task in an independent processor along with the graphical operator interface software. This solution can be costly, however. It is therefore only applicable to systems which can afford a relatively expensive operator interface.

10.1 Graphical Environments

The dominant graphical environments for computers are based on the "windowing" model. This is manifest commercially in the various flavors of Windows, the MacOS, and Motif on UNIX systems. While this is not an ideal environment for mechatronic operator interface implementation it is so common that it is often used for that purpose nonetheless.

Learning windows-style programming has previously been an incredibly intense experience (instructional books typically run around 1,000 pages or more and assume an expert skill level in C or C++); programming windows-style user interfaces has been an incredibly tedious experience. Fortunately, this has changed. This chapter will show the application of easier windowing programming methods to the construction of an operator interface for control.

10.1.1 Windowing Software: Events and Messages

Before proceeding with simplified programming technology for windowed systems a brief overview of how these systems worked will be presented. Since any programming tool that uses windowing must interact with these more basic facilities, this section provides a perspective on what is happening inside the tools to keep the operator interface functioning.

Windowing software differs from previous software models in that it is strictly responsive to events. In prior software models, for example, the kind of program that would be written for numerical analysis, the program is in control. Input from the user, if desired, is solicited by the program. In the windowing model, however, the locus of control passes back and forth between the user and the program. Initially the user is in charge and the dominant paradigm is that the program responds to events generated by the user. However, there are ways for the program to assert control and force the flow of information in much the same way as traditional programs. The primary events are mouse movements, mouse buttons (depress, release), key press, key release. The window management software keeps track of the events, what windows are on the screen and where they are, which programs the windows belong to, which window has the current "focus," where the mouse pointer is on the screen, and so on. It also maintains the display, communicating when necessary to application programs that they must update their parts of the display.

The window management software communicates with user programs by sending them messages. Every event that is relevant to a program generates a message. The primary responsibility of a program is to process those messages, and there are lots of them! Every window controlled by a program can generate messages and the program must keep straight where they came from and what to do with them. Programming for windowing systems at the base level is so tedious because the events represent very low-level activities. In each case, the context of that activity is left for the program to figure out and take the appropriate action.

Much of the action in a windowing program takes place in callback functions. These are user-written functions that execute when specific events take place. A program consists of large numbers of these callbacks. Like real-time programming elements, callbacks execute asynchronously because the events that invoke them are generated unpredictably by the user.

User interaction takes places in two primary domains. The first is the window or group of windows that are unique to an application. This is where the user creates content or views program results. The text input window of a word processor and the drawing window of a graphics program are examples of this type of user interaction. The figure window in Matlab is another example as is Matlab's command-line window. The second type of interaction, the dialog, is more directive. It allows windowing programs to operate temporarily in a mode that is reminiscent of traditional programs in command-line environments (such as DOS). When information is needed from the user, a dialog window is opened in which the user can enter the desired information. Most dialogs are modal, meaning that the program cannot continue until the user completes the information in the dialog and dismisses it. Modeless dialogs, which allow other program activities to continue while the dialog is displayed, are much less frequently used.

10.1.2 Operator Interface vs. Standard Windowing Application

Control system operator interface programs are different from most windowing applications. The primary difference is in the focus of activities. For a standard windowing program, the user is the focus. The raison d'etre of the program is so the user can create content—word processing documents, spreadsheets, data entry, and so forth. In a control system, the target system and the controller are where the focus lies. The operator intervenes from time-to-time to adjust the operation, to set up for a new operation, to shut the machine down, and the like. Most of the time, however, the operator interface software maintains a display that the operator might check occasionally or might monitor to assure proper machine operation.

From this perspective, most of the operator interface looks and behaves like a set of modal dialogs. Only one is present on the screen at a time and until it is dismissed no other display or activity can take place. The operator does not make very heavy use of menus. Menus allow the user to guide the direction of activities whereas in a control system the range of choice for the operator is much more restricted. Some control system operator interfaces don't have any menus at all.

10.1.3 Simplified Programming for Windowing Systems

The complexity of raw windowing programming makes it economically prohibitive for all but high-value applications. It is also a relatively low-productivity environment because so much code has to be written for every aspect of the program.

A number of methods have been developed to make programming for windowing systems easier. Several of these that are suitable for operator interface construction for real-time control programs will be explored here.

10.1.4 The Ease-of-Use Challenge

Before looking at specific solutions to the problem of programming window-based interfaces, we will pose a challenge that can be used to define what "easy" means. The challenge is to write a program that uses a graphic user interface to accept a numerical value from the user, multiply that value by 2, and then display the result in the same window. Although simple, this will at least give a common ground on which to compare various approaches.

10.1.5 Methods of Simplifying Window-Style Programming

One of these, the use of a terminal window, has already been introduced. The character-based operator interface described in the previous chapter can run in a windowing system if it supports a terminal-mode window. Such a window behaves exactly like the command-line screen of traditional operating systems thereby allowing standard programs to run (for example, C++ programs using scanf()/printf() or cin/cout for user input and output). The character-based operator interface is constructed by using cursor control in that environment to get a nonblocking user interface that is highly portable.

Two other programming environments will be explored, Visual Basic (from Microsoft) and Bridgeview (from National Instruments), a close relative of Labview. These represent two different approaches to graphic operator interface construction. Visual Basic is more complex and more general. Bridgeview is specialized for operator interface and other programming needs associated with industrial control whereas Visual Basic is a general programming language.

10.2 The Times-2 Problem

Although simple, this problem really brings out the fundamental characteristics of user interface construction methods. The three examples that will be shown here highlight three very different methods of dealing with operator interfaces. The character-based interface uses classical programming methods to create a nonblocking means of data entry and display. It creates an ad-hoc version of the event-based software that is the more modern approach used in standard windowing systems. Visual Basic exposes the full windowing programming paradigm, but provides very strong assistance to the programmer in working within that model so that the amount of work needed is reasonable. Bridgeview operates within the windowing environment, but completely hides it from the programmer.

10.2.1 Times-2: Character-Based Interface

The operation of the character-based operator interface has already been explained in detail. The code in this example shows the minimum necessary to do the times-2 function. The first part does the setup of the screen elements. The while() loop keeps the screen refreshed until the user presses Ctrl+X to exit.

Section 10.2. The Times-2 Problem

```
main ()
   {
   double A_Double = 0.0;
   double x2;

   COperatorWindow* TheOpWin = new COperatorWindow ("Times-2 Test Window");

   TheOpWin->AddInputItem ("A Double:", &A_Double);
   TheOpWin->AddOutputItem ("Times 2", &x2);
   TheOpWin->AddKey ('X', "Stop", "Program", DumpOut);

   cout << "Maximize window if necessary. ENTER to continue...\n";
   getch();

   TheOpWin->Display ();

   while(DoExitNow == 0)
       {
       TheOpWin->Update ();
       x2 = 2.0 * A_Double;
       }

   TheOpWin->Close ();
   clrscr (); // Clear the operator interface
   // Delete objects that were created with 'new'
   delete TheOpWin;

   cout << "Hit ENTER to exit.\n";
   getch();
   return (0);
   }
```

Very little code is required and very little training is needed to use this operator interface software. There are only three types of screen items: input items, output items, and action keys. Each one has a setup function call—the data type of the input or output items are detected internally through the use of function overloading. Once the setup has been accomplished, all of the screen items are updated automatically; only the associated variable(s) need to be computed to change the screen display of the output.

The screen display is shown is shown in figure 10-1 (spaces have been removed for compact display).

As noted earlier, the screen display certainly does not meet contemporary standards for user interfaces. However, it is nonblocking, fast, portable, and about as easy as can be imagined to program.

```
       Times-2 Test Window

A Double:       ->      1.34e-05        Times 2      2.68e-05



Ctrl-X
Stop
Program
```

Figure 10-1. Times-2 Test: character-based operator interface

10.2.2 Times-2: Visual Basic

Visual Basic revolutionized Windows programming by making it accessible to casual programmers, those who use programming as part of their activities but do not do it full time.[1] The complexity of the raw windowing application interface and the amount of learning time needed to become proficient in it acted as strong barriers to this class of programmers. Visual Basic changed that. Not by changing or hiding the windows programming model, but by making that model interactive. The visual part of Visual Basic allows for creation of screen items interactively, using the mouse to position and size various objects. The next step is the important one. Instead of relying on memory and a user's manual to figure out what events will be generated from each kind of object, then writing the code to respond to those events, Visual Basic presents lists of possible events for each object you have created and then generates the core code needed to respond to that event. The programmer is left only with the task of filling in the specifics of the response. Thus, the windowing programming paradigm is left intact but the programmer only has to understand the principles since the detail information is presented as needed. On the other hand, issues such as data type for numeric data input is still left to the user resulting in more work for the programmer than in the character-based interface.

Figure 10-2 shows the window produced using Visual Basic for the times-2 problem. Numbers are entered into the upper edit box (labeled x) and the result is shown in the lower box. Each time a new number is entered its doubled value appears automatically. The user signals completion of an entry by either typing Enter or using the mouse or keyboard to remove the focus from the entry window.

Creating this program starts with use of the graphic input to add the two edit boxes and their associated labels to the window (a form). Visual Basic then creates the code for the function that will execute with the creation of the window. The function created is:

```
Private Sub Form_Load()
    Mult2
End Sub
```

---

[1] archive: grp_cpp.zip, directory: VisualBasic/Times2

Section 10.2. The Times-2 Problem   161

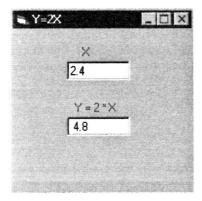

Figure 10-2. Times-2 using Visual Basic

The middle line is user-added; Visual Basic creates the empty function and arranges that it will be called when this window (form) is loaded. The programmer does not have to worry about the details of how the function got called. The function mult2 is:

```
Private Sub Mult2()
    xx = Val(x.Text)
    y = 2 * x
    y2x.Text = Str(y)
End Sub
```

This function is where the main conversion work is done. The conversion is necessary because Visual Basic only recognizes text in edit boxes. Interpretation of the text as a number is up to the programmer. The key to this code is that the edit boxes have been named (by the programmer) x and y2x. Each box has associated properties which are accessed in the program using the dot notation of data structures. Thus, x.Text accesses the current text associated with the edit box. The function Val() converts that text to a number. The last line converts the newly calculated number back to text for storage into the Text property of the edit box y2x. Note that data conversions of this sort are handled automatically in the character-based user interface.

The program is not yet complete, however. Something must be done to cause the function Mult2 to execute when the user types a new value into the input edit control. It is up to the programmer to decide on the events that will cause this to happen. The most reasonable are a loss of focus event and the user pressing Enter. Loss of focus, as noted, happens whenever the user does something that causes the windowing manager to stop using the particular control element as the place where new keystrokes are placed.

Again, Visual Basic takes care of the details of setting up functions to respond to these events. The function which follows is created when the programmer selects the event "key press." However, the programmer must fill in the code to detect which key triggers the action (the value `vbKeyReturn` is a predefined constant) and what to do if that key is detected. Note that this function is called for every key press but does something only if the selected key is detected.

```
Private Sub x_KeyPress(KeyAscii As Integer)
    If KeyAscii = vbKeyReturn Then
        Mult2
    End If
End Sub
```

The function that responds to the loss of focus event is similar, but simpler because no selection has to be made; it always runs `Mult2`.

```
Private Sub x_LostFocus()
    Mult2
End Sub
```

The conclusion? Visual Basic provides a reasonable way to implement the times-2 test, but it requires considerably more understanding of the windowing programming model than is required by the character-based operator interface. It is, however, capable of producing a full windowing operator interface, unlike the character-based system.

10.2.3 Times-2: Bridgeview

Bridgeview takes a very different approach than either the character- based interface or Visual Basic.[2] Like Visual Basic, it uses interactive graphical design to put items on a window that will become the operator interface. The programmer then uses a graphical programming language (called G) to connect those elements. The choice of elements is much richer than it is in Visual Basic; the elements also include data type information so that conversions are handled internally rather than by the programmer and have knobs, sliders, and so forth, for graphical representation of data.

The operator interface window (or panel as it is called in Bridgeview) as created using Bridgeview is shown in figure 10-3. Input can be entered in one of three ways: by typing a number into the edit box, by using the up/down arrows on the edit box to change the value, and by grabbing the knob with the mouse and moving it to the desired value. As the input is changed, the output will follow, displayed as a colored bar and as a numeric value.

The elements are placed from menus of available elements and identified as either input (control) or output (indicator). The programmer must supply the connection between these items, using G, a graphical, data flow language. It uses lines that resemble wires to connect the elements on the user interface window. They are not really wires, of course. They are data flow channels. A data flow channel transmits a packet of information (numerical in most of our cases) from a data source (an

---

[2]archive: `grp_cpp.zip`, directory: `Bridgeview/Samples`

Section 10.2. The Times-2 Problem 163

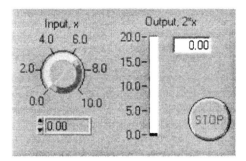

Figure 10-3. Times-2 using Bridgeview

input or control item) to a data consumer (an output, or indicator, in the times-2 case). Each time a new value is generated by the source a new packet is created and transmitted to the consumer. The consumer processes its input packets when it has a complete set of input packets present (that is, each of its input lines has a packet waiting). It doesn't do anything until its input packet set is complete.

The G program for the times-2 problem is shown in figure 10-4. The lines connecting the elements are the data flow channels. Each of the elements is the programming double of the visual element on the user interface screen. They are placed on the programming screen (the diagram window) automatically by Bridgeview at the time they are created by the programmer.

This program has three visual elements, corresponding to the user input, the output, and the stop button. All of the items in the programming window are color coded to show data type. Note that the control and indicator items also have the data type shown as part of the item (DBL for double, TF for true/false logic). It also has a number of processing elements that have been placed on the program diagram but do not have visual counterparts. Before going over the details of the times-2 program, we will look at a couple of simple examples to illustrate the principles of G programming.

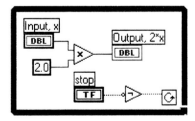

Figure 10-4. Times-2 using Bridgeview: G program

The first example computes the result of the quadratic polynomial

$$y = x^2 + 2.5x + 1.5$$

For each value of x entered, the equivalent value of y is displayed. The program executes once. After a new x value is entered, the program must be executed again for the new value of y to appear. Figure 10-5 shows the panel for the quadratic after a value has been entered and the result computed.

This example gives a good opportunity to illustrate the data flow rules of computation. Diagrams consist of sources, processors, consumers, and data-flow wires. The program diagram is shown in Figure 10-6. A program such as this is called a "virtual instrument," or VI. The term comes from the Labview usage which is targeted primarily to the data acquisition community. A complete system can be made up of many VIs.

When execution is started, each data source produces a packet of information. Unless a control structure is used (which is not the case here) the source will only produce one packet per execution.

This diagram has three sources, the control labeled x which shows on the panel, and the two constants, 2.5 and 1.5. When execution starts, each of them generates a data packet. The packet from the 2.5 constant item travels to the lower multiplication processor. The multiplication processor (shown as an x in a triangle), however, does not do anything until packets are present on both of its inputs. Meanwhile, the packet from the x-control travels to the dot, indicating a split in the wire. A split causes packet duplication, i.e., the same packet is sent down both legs of a split (there is no conservation law for packets). When the packet from x arrives at the multiplier it will have packets at both of its inputs and will perform the indicated operation. When the multiplication is complete, a new packet is formed and launched toward the summing processor (shown as a rectangle with a + and three input boxes). This process continues until all of the channels are clear and the indicator (y) has displayed the new value, after which Execution halts. (Describing this in words make the process sound very laborious, but it actually all happens quite efficiently.)

The general rules for data flow execution are:

- Each source generates a data packet at the start of execution

- Processors execute when all of their inputs have packets

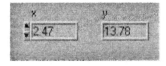

Figure 10-5. Bridgeview panel for quadratic

Section 10.2. The Times-2 Problem   165

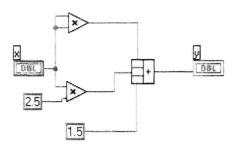

Figure 10-6. Quadratic—G program

- Consumers (indicators) execute when their inputs are all filled

- Execution halts when no new packets are being generated and all channels are clear

This pure data flow paradigm, however, does not allow for the solution of many interesting problems. It was originally designed for hardware-based processing units in which there is a continuous flow of input data. In its adaptation as a graphical computing language something must be added to provide the equivalent of continuous input flow of data. In G, these are looping control structures, the graphical equivalents of for and while loop constructs in algorithmic languages. Other constructs have also been added for switch and sequence control.

Figure 10-7 shows a repeat of the times-2 Bridgeview program. The wide line (in gray on screen) surrounding the items is a while loop. Its rule is that it continues executing as long as the signal sent to the box in the lower right corner is true. "Continuing" execution means that the for each loop of the while, the source elements send out new data packets. This keeps the execution alive as long as the loop continues. When the stop button is pushed, the signal to the while loop control box becomes false and execution ends (the processor between the button and the loop control is a logic inverter). Note that there are two disconnected data flow

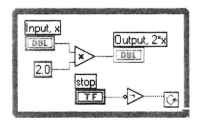

Figure 10-7. Times-2 using Bridgeview: G program (repeated)

paths within the while loop. These execute (nominally) in parallel. The data flow rules apply equally to parallel paths as they do to a single path.

Although the description of the Bridgeview times-2 program took a while because of its considerable difference from conventional programming, for a person who has learned G creating this application is very simple. Of the three methods described here, character-based, Visual Basic, and Bridgeview, Bridgeview is the simplest and fastest to use. A modestly experienced G programmer can create this application in less than five minutes. The character-based is next easiest followed by Visual Basic (which is still enormously easier than doing the same thing using raw windowing program methods).

10.3 Screen Change

One of the features that is necessary in an operator interface is the ability of the software to control the change of display that the operator sees. This happens as the system goes through various stages of operation or as the operator requests display of different information. A general rule for these display changes is that they should never surprise the operator. Any change should be at the direct request of the operator or with substantial warning to the operator. A screen change is an asynchronous operation so if the operator is doing something on the screen at the time the change takes place, there can be uncertainty as to what action the system has taken. The only exception to this is for emergency situations, serious alarms, and so on.

10.3.1 Screen Change in Visual Basic

The screen change test program uses a master screen for the user to select which screen to display.[3] This function would normally be done internally in a control program but since there is no control program in this sample, it is left for the user to do.

The user selects which screen to display by number. The two screens are the times-2 screen from the previous example, and a second screen that shows a simple graphics operation, moving a user-selected object on the screen. The master software manipulates which screen is displayed by changing the Visible property of the relevant window (form). The code is in the master program,

```
Private Sub SelectScreen()
    scrn = Val(ScreenNumber.Text)

    ' Hide previous screen
    Select Case prevscrn
        Case 1
            Screen1.Visible = False
        Case 2
            MoveObject.Visible = False
    End Select
```

---

[3]archive: **grp_cpp.zip**, directory: **VisualBasic/ScreenChange**

Section 10.3. Screen Change

```
' Display current screen
Select Case scrn
    Case 1
        Screen1.Visible = True
    Case 2
        MoveObject.Visible = True
End Select

prevscrn = scrn   'Remember current screen
End Sub
```

The user input is handled in exactly the same way the user input was handled in the times-2 program.

The second screen, MoveObject, uses some basic graphics commands to move a user-selected shape on the form. The appearance of the screen is shown in figure 10-8.

This program demonstrates more of the capability of Visual Basic to create the kinds of screens that are needed for control system operator interfaces. Moving on-screen objects is often done to animate the behavior of the target system so the operator can see what it is supposed to be doing. In systems for which the operator cannot get a good view of the operation the animation serves as a stand-in. Not being able to see the whole system is very common, due to operation inside an environmentally controlled space (oven, vacuum, etc.), because of intervening objects that block the view, because of safety considerations, and so forth.

The code to shape and move the object makes use of a number of properties. The power of Visual Basic is that these properties are easily accessible to the programmer by picking them from lists that appear as different objects are selected on the graphical display of the form being designed.

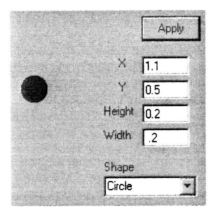

Figure 10-8. Move-Object screen: Visual Basic

```
Private Sub PlaceObject()
    x = Val(Pos_x.Text) * 2000
    y = 2000 - Val(Pos_y.Text) * 2000
    ht = Val(ObjHeight.Text) * 2000
    wdth = Val(ObjWidth.Text) * 2000
    ShapeIndex = ShapeType.ListIndex
    Shape1.Shape = ShapeIndex
    Shape1.Left = x
    Shape1.Top = y

    Select Case ShapeIndex
        Case 1, 3, 5 'Square, circle, rounded square
            ' These shapes have only one
            ' dimension - height and width are set equal
            ' to the "height" as entered on the form
            Shape1.Height = ht
            Shape1.Width = ht
            ObjWidth.Text = Str(ht / 2000)
        Case 0, 2, 4 'Rectangle, oval, rounded rectangle
            Shape1.Height = ht
            Shape1.Width = wdth
    End Select
End Sub
```

10.3.2 Screen Change: Bridgeview

Screen selection in Bridgeview is accomplished by manipulating VIs. VI is the Bridgeview equivalent of a subroutine. Unlike an ordinary subroutine, however, a VI has an associated window (panel) for display of its active interface in addition to the window for its G-code representation. Screen selection is done by manipulating the properties of these panels, so that a panel is only displayed when its associated VI is actively executing. Thus, turning the various screens on and off is just a matter of calling the right VI.

The structure of the Bridgeview program parallels that of the Visual Basic program—there is a master screen for the user to pick which screen to display from a choice of two. One of these is again the times-2 program. The other is a similar program (it divides by 2) but uses sliders as graphical elements for both input and output. The master screen is shown in figure 10-9, the screen using sliders for input and output in figure 10-10.

Values on the input (left) side can be changed by grabbing the pointer with the mouse and sliding it to the desired value. The numeric value is shown as the slider is moved. Alternatively, an input number can be entered in the text box.

Figure 10-11 shows the G program to implement the screen change. The program actually consists of three panels, but all are shown here for illustrative purposes. The box that looks like a strip of film is a sequence control. It forces the execution to follow the indicated sequence, shown by the sequence number at the top of each frame. When programming on screen, only one step of the sequence is seen at a time. The first step in the sequence (frame #0) loads the relevant VIs into memory so they are available for immediate execution. The second step reads the user input

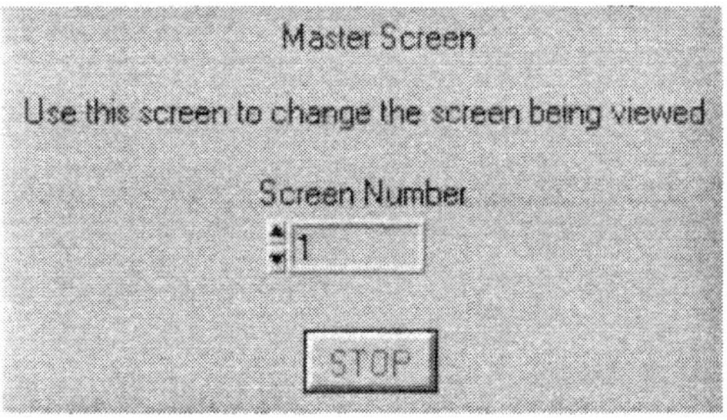

Section 10.3. Screen Change   169

Figure 10-9. Master screen; Bridgeview screen change

Figure 10-10. Slider screen; Bridgeview screen change

and sends the value to the screen selection VI. The small arrows along the side of the interior while loop are called shift registers. They are used to save values from one iteration to another. The saved value is used in the screen selection program to know when to close the previous window.

A single view of the screen selection VI is shown in figure 10-12. Since a number of multiview elements are used, it is difficult to see the whole program in printed form, although it is very easy to go through it on screen. The particular view shown has an outer while loop with inner controls for true/false, sequence, and selector. It controls closing of the previous panel (window) so that a new one can be opened. While this VI has a panel, as do all VIs, it is never shown. The input and output items on that panel become the VI's arguments and are passed in from the calling VI. Because Bridgeview does not have any equivalent to pointers or other ways of making lists of references to VIs, each screen to be selected must be programmed explicitly by using a selector control.

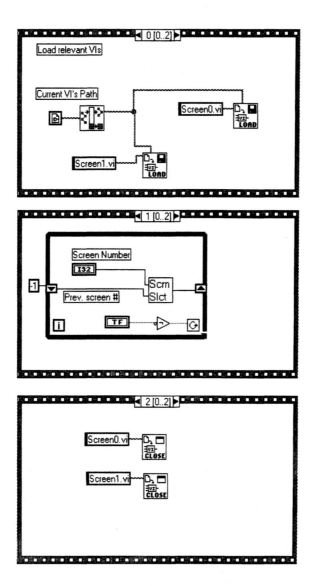

Figure 10-11. G Program for screen change

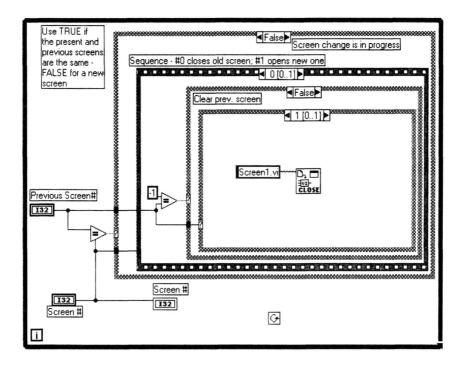

Figure 10-12. Screen Selection VI

10.4 Heat Exchanger Control in Bridgeview

As an exercise to prepare for the use of Bridgeview with C++ control code a heater control example has been fully programmed in Bridgeview.[4] Although it is not intended to be used as an actual control program, if the target system were slow enough it could be used for that purpose.

The program uses two screens, one for setup (figure 10-13) and one for operation. It is essentially the same as the screen that will be used when Bridgeview is used for the operator interface to the C++ control code. It allows for setting controller gains and specifying how long the simulation should run. The control buttons provide for the start of operation mode or for terminating execution of the program.

The simulation itself is a fairly elaborate program that includes definition of a PID controller as well as a Euler integration of the system equations for the simulated thermal system. Because of the extensive nesting used in the program, it is not worthwhile to try reproducing it here. It can be easily examined on screen, however.

The operation panel (window), shown in figure 10-14, shows the temperature and desired temperature as well as a graph showing the evolution of temperature with time.

---

[4]archive: `grp_cpp.zip`, directory: `Bridgeview/HeaterBV`

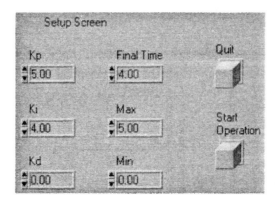

Figure 10-13. Heat control in Bridgeview: setup screen

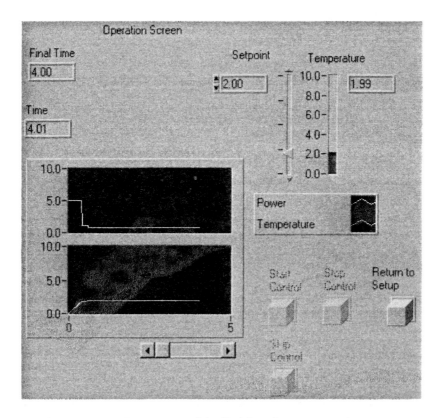

Figure 10-14. Heat control in Bridgeview: operation screen

The setpoint can be changed using either the numeric input or the slider. Graphic output displays of this sort are critical for many control systems that evolve reasonably slowly over time. The operator can watch such a display and have a very good idea whether or not the system is operating properly.

The control buttons allow for starting and stopping the control or returning to the setup mode. The lower button, Skip Control, provides a means to get back to the setup screen without starting the simulation. This is very useful when it is only realized after entering the operation mode that some setup parameter must be changed. Note also that some of the buttons appear grayed. These are nonoperative buttons. They are made operative internally when the system is in appropriate states for those buttons to be active.

10.5 Interprocess Communication: DDE

When one of these operator interface methods is used, the operator interface will usually run as a separate program, distinct from the C++ control program (or the portion of the control program that is running in the computer that is used for the operator interface). Although this is not absolutely necessary, it is more convenient to keep them as separate entities than to try to use various multi-language facilities to combine them into a single program. Separate programs represent independent processes in the terminology used here. Independent processes do not share an address space so interprocess communication must be done explicitly rather than using language syntax. All of the operator interface programs used here run in Windows so a communication method appropriate to Windows will be used. Interprocess communication is one of those areas where it is difficult to find portable techniques.

Of the several methods for interprocess communication that Windows supports we will use one of the simplest and oldest, dynamic data exchange (DDE). This protocol only allows for exchange of character strings so is not a particularly efficient way to exchange numeric information. On the other hand, it is available very widely and is easily adapted to our needs.

DDE is a client/server architecture in that one partner in the exchange is normally expected to initiate exchanges by asking information (the client) while the other (the server) supplies the requested information. As with our usage of TCP/IP, which is also set up as a client/server architecture, we will use DDE in much more of a peer-to-peer arrangement as the operator interface program and the control program exchange information.

10.5.1 DDE: The C++ Side

While the details of implementing the DDE protocol can get quite complicated, a C++ user interface can be abstracted from it which is quite useful for our needs and simple to use. A set of test programs demonstrating DDE usage is in the Bridgeview/DDE-Test directory of the grp_cpp.zip archive. Its only function is to show that information can be exchanged between programs. It sets up a DDE link and then uses the character-based operator interface to demonstrate data exchange of various data types.

Both client and server examples are given. The client sample program (CDDE_Client.cpp) is set up so that it can communicate either with the sample server program or with Microsoft Excel.

The first step of the DDE setup is to instantiate a DDE entity:

```
// Create a DDE client object
MyClient = new C_DDE_Client ();
```

Then, the topic(s) must be defined. In this case, only one topic is used,

```
if(TalkTo == TALK_TO_CONSOLEDDE)
   Inputs_Topic = MyClient->AddTopic ("Server Data");
else if(TalkTo == TALK_TO_EXCEL)
   Inputs_Topic = MyClient->AddTopic ("Sheet1");
```

The name used for the topic must match the name defined in the server. "Sheet1" is the name that Excel uses as the default for the first spreadsheet page.

The item specification connects the DDE channel to specific C++ variables. Each variable is associated with an item. Again, the item names used in the client and the server must match.

```
if(TalkTo == TALK_TO_CONSOLEDDE)
   {
   LFC = Inputs_Topic->AddItem ("LongFromClient", LongFromClient);
   DFC = Inputs_Topic->AddItem ("DoubleFromClient", DoubleFromClient);

   // These are the items which will be server outputs and client inputs
   IFS = Inputs_Topic->AddItem ("IntFromServer", IntFromServer);
   FFS = Inputs_Topic->AddItem ("FloatFromServer", FloatFromServer);
   SFS = Inputs_Topic->AddItem ("StringFromServer", StringFromServer, 32);
   CNT = Inputs_Topic->AddItem ("Counter", Counter);
   }
else if(TalkTo == TALK_TO_EXCEL)
   {
   LFC = Inputs_Topic->AddItem ("R1C1", LongFromClient);
   DFC = Inputs_Topic->AddItem ("R2C1", DoubleFromClient);

   // These are the items which will be server outputs and client inputs
   IFS = Inputs_Topic->AddItem ("R1C2", IntFromServer);
   FFS = Inputs_Topic->AddItem ("R2C2", FloatFromServer);
   SFS = Inputs_Topic->AddItem ("R3C2", StringFromServer, 32);
   CNT = Inputs_Topic->AddItem ("R4C2", Counter);
   }
```

The first part, for connection to the C++ DDE server program, uses names that match the usage in the server. The second part, connecting to Excel, uses the names defined in Excel to connect data items with the rows and columns of the spreadsheet—R3C2 is for row 3, column 2. The variables must persist after the function containing these statements returns because the actual connections

Section 10.5. Interprocess Communication: DDE

(reading and writing to these variables) is done elsewhere. The scheme here is very similar to that used for the character-based operator interface except that references are used here and pointers are used in the character-based operator interface.

The lines following this section are needed because of the peculiarities of the client/server structure. Recall that in client-server architectures, the server is normally passive. It serves by waiting for a request from a client and then responding. Many DDE applications (as well as other client-server technologies) require more of a peer-to-peer architecture. To achieve that, DDE includes the advise facility so that the server can be just active enough to satisfy these needs. When the client sets up one of its items in advise mode, whenever the server detects that the variable has changed it will send a message to the client saying, "please ask me about XXX now because it has changed." The client can then ask to complete an interchange initiated by the server (even though servers aren't supposed to initiate anything).

```
IFS->AdviseLink ();
FFS->AdviseLink ();
SFS->AdviseLink ();
CNT->AdviseLink ();
```

The final step is to make the connection. The server name and topic name are required to make a DDE connection. Excel uses its program name as its server name. The console-dde server program uses the name "Testing." The topic names were defined earlier.

```
// Connect to server and start advise links to update changing data
if(TalkTo == TALK_TO_CONSOLEDDE)
   MyClient->ConnectToServer ("Testing", "Server Data(";
else if(TalkTo == TALK_TO_EXCEL)
   MyClient->ConnectToServer ("Excel", "Sheet1(";
```

Program execution involves getting data sent by the server (yes, it is technically true that the server didn't really send the data, but the C++ programming interface is designed to hide that fact) and sending data to the server. Notice, again, that this is very much a peer-to-peer arrangement. While the client does send data to the server, there is no response expected.

GetValue() gets data values from the server. The while() structure is because the internal structure of these DDE functions uses a separate operating system thread to actually service the DDE channel. If the data being requested is in the process of being changed, GetValue() will refuse the request.

```
while (!CNT->GetValue (Counter));  // Get data changed by the server
while (!IFS->GetValue (IntFromServer));
while (!FFS->GetValue (FloatFromServer));
while (!SFS->GetValue (StringFromServer));
```

Sending values from the client to the server is similar. One extra step is inserted, however, in the interests of efficiency. That step is to check to see whether the variable in question has changed. If it has not, there is no need to send it. Sending data is an expensive process (in the computing resource sense) so this extra step usually pays off. When the variable changes constantly, a change delta is often defined so that only changes that exceed that incremental value are sent. The same data integrity problem exists with sending as with receiving, so the same while loop construct is used.

```
if (LongFromClient != Last_LFC)
    {
    while (!LFC->SetValue (LongFromClient));
    Last_LFC = LongFromClient;
    }
if (fabs (DoubleFromClient - Last_DFC) > 1E-6)
    {
    while (!DFC->SetValue (DoubleFromClient));
    Last_DFC = DoubleFromClient;
    }
}
```

The last step is to disconnect when the program has finished,

```
MyClient->DisconnectFromServer ();
```

10.5.2 Communicating with Excel

With the TalkTo variable set to TALK_TO_EXCEL the cdde_client.cpp program will compile using the proper DDE names to exchange information with the Excel spreadsheet program. When the program is run, values entered from the C++ program will appear in spreadsheet cells and values entered into appropriate spreadsheet cells will appear in the operator interface window of the C++ program.

Program output is shown in figures 10-15 and 10-16. The character operator interface display has had blank spaces removed to save space. Numbers in the left column of figure 10-15 are entered by the user and transmitted to Excel, where they appear in the left column of figure 10-16. Data entered in column B of the spreadsheet appear in the righthand column of the C++ program.

A bug (or feature!) connected with this transfer should be noted. When Excel sends a string via DDE it appends additional characters, one of which is a carriage return (\r). When this string is displayed in the C++ program, it causes the cursor to move to the extreme left of the screen and puts the extra blanks added to ensure that old information is erased into the empty space there. Thus, if a string that is shorter than the string it replaces is transmitted, part of the old string remains on the screen. But, is this a bug or a feature? If it is interpreted as a bug, extra code could be added to remove those characters. On the other hand, the DDE client program is general. If characters were removed arbitrarily, the program might malfunction when used in an environment for which those characters had meaning. In this case, the decision was made to leave the characters as transmitted and note the display behavior as a peculiarity of communicating with Excel.

Section 10.5. Interprocess Communication: DDE   177

```
DDE Client Test Program

Long(R1C1):        4321583        Integer(R1C2): -54
Double(R2C1):  ->  0.0832         Float(R2C2):   1.2e+07
                                  String(R3C2):  Excel
                                  Counter(R4C2): 9742




Ctrl-X
Exit
Program
```

Figure 10-15. DDE comunication with Excel: character-based operator interface

	A	B
1	4321583	-54
2	0.0832	1.20E+07
3		Excel
4		9742
5		
6		

Figure 10-16. DDE comunication with Excel: spreadsheet output

10.5.3 A DDE Server in C++

Using the DDE server in C++ is almost the same as using the DDE client. This is consistent with the applications we are trying to build in the sense that these applications are really peer-to-peer rather than client/server so the usage should be symmetric.

The `console_dde.cpp` program implements a test of the DDE server. Setting the TalkTo variable to TALK_TO_CONSOLEDDE in the client test program (the same program that was used for talking with Excel) causes the program to execute using DDE names that are consistent with the server DDE program. Both programs use the character-based operator interface. Figures 10-17 and 10-18 show the screens for the client and server programs respectively. In each case, the left column is for user input and the right column is for displayed values.

```
 DDE Client Test Program
Long(R1C1):        555          Integer(R1C2):  9876
Double(R2C1):  ->  3.4329e+09    Float(R2C2):):  8e-06
                                 String(R3C2):   New String
                                 Counter(R4C2):  267000




Ctrl-X
Exit
Program
```

Figure 10-17. DDE communication: client C++ program

```
 DDE Server Test Program
Integer:       9876              Counter:  335369
Float:         7.76e-06          Long:     555
String:    ->  New String        Double:   3.4329e+09




Ctrl-X
Exit
Program
```

Figure 10-18. DDE communication: server C++ program

10.5.4 DDE Communication Between C++ and Visual Basic

Items in Visual Basic forms can be set up as DDE clients by specifying the names of the server, topic, and item.[5] To do this, a text box is set up on a form and then the Link properties of the text box are set in the properties table. The section of the properties table for one of the items in this program is shown in Figure 10-19. The server and topic names are specified in the LinkTopic property and the item name is given in the LinkItem property. The LinkMode is left at 0. It will be set in the code associated with setup of the form.

The Form_Load function in Visual Basic executes when the form is first accessed as the program loads. For this program, the only initialization needed is to set the modes for the DDE items. The items that are sent from the client to the server

---

[5]The Visual Basic DDE test program is in the VisualBasic\DDETest directory of the grp_cpp.zip archive.

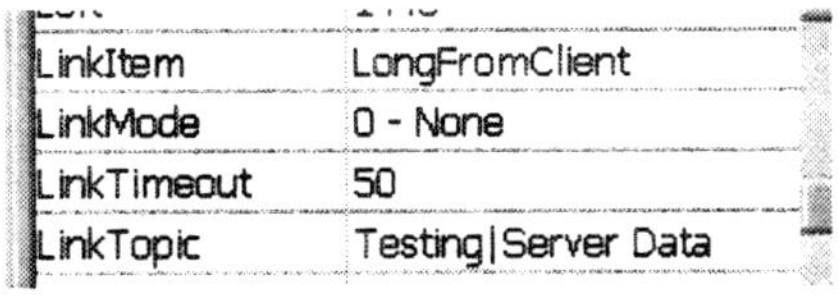

Section 10.5. Interprocess Communication: DDE

LinkItem	LongFromClient
LinkMode	0 - None
LinkTimeout	50
LinkTopic	Testing|Server Data

Figure 10-19. DDE setup for item in Visual Basic

are set to manual mode (mode 2) and the values that are sent from the server to the client are set to automatic mode (mode 1). The initialization section of the Form_Load() is:

```
Private Sub Form_Load()
    On Error GoTo NoSource

    LongFromClient.LinkMode = 2
    DoubleFromClient.LinkMode = 2
    FloatFromServer.LinkMode = 1
    IntFromServer.LinkMode = 1
    StringFromServer.LinkMode = 1
    Counter.LinkMode = 1
    Exit Sub
    ...
```

For variables in manual mode the LinkPoke operation must be used to initiate the transmission. The code which follows shows a section where a change in one of the client-originated variables has been ordered by the user (detected in this case either because the user has clicked the Enter (Return) key or by loss of focus by doing a mouse click outside the text box). The information from the DoubleFromClient text box is sent for DDE transmission by execution of the DoubleFromClient.LinkPoke statement.

```
Private Sub DoubleFromClient_KeyPress(KeyAscii As Integer)
    If KeyAscii = vbKeyReturn Then
        KeyAscii = 0
        DoubleFromClient.LinkPoke
    End If
End Sub

Private Sub DoubleFromClient_LostFocus()
    DoubleFromClient.LinkPoke
End Sub
```

Figure 10-20 shows the Visual Basic screen when this program is run with the C++ DDE server program. No change is required to the C++ program as discussed earlier. As long as a compatible DDE client is running it can communicate.

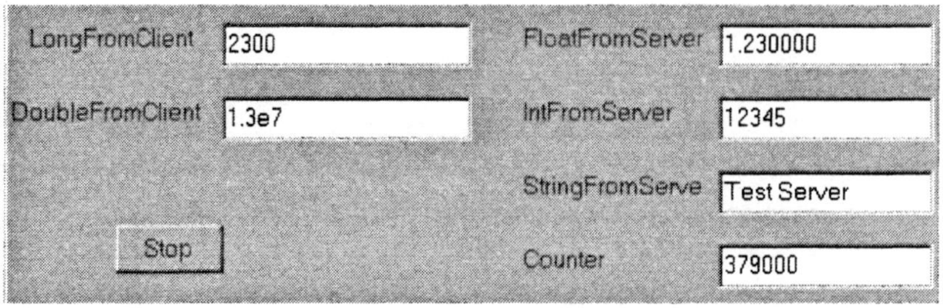

Figure 10-20. DDE test screen—Visual Basic

10.5.5 DDE Communication Between C++ and Bridgeview

Bridgeview is set up to operate as a DDE client so the DDE server program is used for the C++ part of the communication. The tag structure of Bridgeview is used for this purpose. Each DDE data item is configured as a tag, which can then be accessed very easily in G code. The G code for several of the tags that connect to the sample DDE server program is shown in figure 10-21. The tag loops with the funny symbol in the upper right have been completely set up by the tag wizard. The others used the tag wizard initially and then were edited. The metronome-like icons are timers that control how often tag values will be sampled.

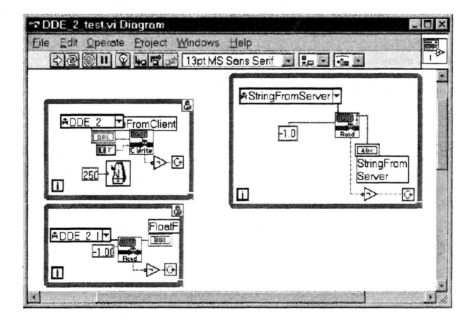

Figure 10-21. DDE tags in Bridgeview

Setting up the tags is done with the Bridgeview Tag Configuration Editor. The server name, topic name, and item name are entered for each tag and the tag is defined as input or output. Other standard tag parameters control amount of change before new values are sent, alarm limits, and so forth (see the Bridgeview manual for details on the Tag Configuration Editor). The file dma1.scf saves the tag configuration information.

The program operates in much the same manner as the previous tests. A screen capture of the Bridgeview panel is in figure 10-22. It runs with the C++ DDE server program described earlier. As with the Visual Basic example, the C++ program is the same as in the previous tests. The items on the left are input from the Bridgeview side; those on the right come from the C++ server.

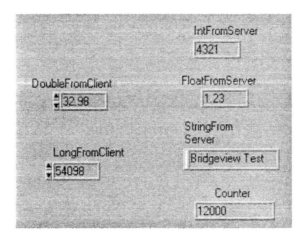

Figure 10-22. Bridgeview screen for DDE test

10.6 Putting It All Together

All of the pieces are now in place to get a graphically based operator interface to work with a standard C++ task/state based control program.[6] The operator interface task has to be rewritten to use DDE to communicate with the program managing the GUI (graphical user interface) and the Bridgeview or Visual Basic programs have to be written, all using the DDE technology.

The operator interface task of the C++ program (Heat_op.cpp) has been rewritten so that it can support either a DDE-based operator interface or a standard, character-based operator interface. One of the #include symbols, USE_CHAR_OP or USE_DDE, must be defined to set the mode.

---

[6] The C++ programs and the Bridgeview programs are in the BridgeView\HeaterDDE directory and the Visual Basic programs are in the VisualBasic\HeaterVB directory of the grp_cpp.zip archive.

Each state represents a screen, in both the character and graphic (DDE) operator interface modes. The DDE mode uses a separate variable for screen number which is transmitted to the program managing the graphic mode interface.

The Visual Basic and Bridgeview programs combine the screen change material and the DDE communication test program material. These are applied in a straightforward manner, with very little change from the test programs except for variable names.

Figures 10-23 and 10-24 show the resulting GUIs. The Visual Basic screen (figure 10-23) at this stage is little more than a graphical rendition of the character-based operator interface. A wide variety of plug-in Active-X or OCX modules are available for Visual Basic to allow construction of more sophisticated screens. On the other hand, Bridgeview's built-in elements allow for graphic as well as numeric display of system information (figure 10-24).

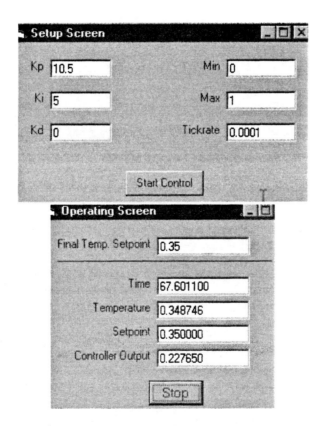

Figure 10-23. Visual Basic: setup and operation screens for heater control

Section 10.6. Putting It All Together

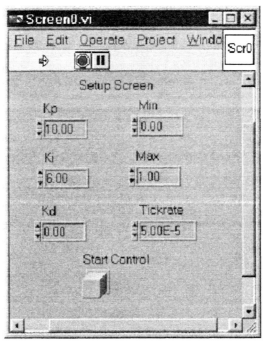

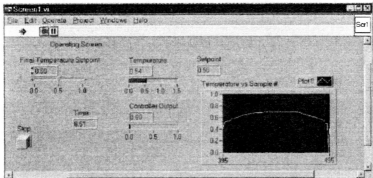

Figure 10-24. Bridgeview setup and operation screens for heater control

Chapter 11

DISTRIBUTED CONTROL I: NET BASICS

Control activities can be distributed across several computers for several reasons—the most common is to match the geographic distribution of the control object. In this case, wiring to instruments and actuators is reduced by placing the control computer close to them. Only network wiring is then needed to connect the control node to the rest of the control system. This arrangement can be a major cost saver in a complex system. Probably the next most likely reason to distribute control is when latency requirements for several tasks cannot be met with a processor of reasonable cost. Each of these tasks can be placed in a processor in which it is the highest priority, thereby meeting all of the latency requirements.

The performance price paid for distribution is in communication. Information that must be shared among tasks now must pass across some communication medium. This must be compared to the fairly low performance cost associated with guarding information integrity, as described in chapter 6 (or no cost at all for a system restricted to single thread operation). The communication performance price is why only a very few multiprocessor configurations can be used to actually increase the total amount of computing resource available to the job.

Several potentially useful communication arrangements will be described in this chapter. Of these, a network configuration using TCP/IP protocols will be used for detailed examples. The examples in this chapter will deal only with how to move information around on the network. Chapter 12 will show how to use these techniques for system control.

The final problem associated with distributed control is how to partition a uniprocessor program into multiple chunks. No partitioning will be used here except the natural partition associated with the task structure. Thus, each processor in the distributed system will execute the code for one or more tasks. Further distribution is theoretically possible. However, the same effect can be realized by splitting a compute-heavy task into two or more smaller tasks.

11.1 Multiprocessor Architectures

The architectures described here range from very tightly coupled structures with very fast data transmission rates, to very loosely coupled systems with substantial delays for data exchange. A combination of data rate, geographic constraints, and cost will govern actual choice of architecture for a specific control application.

11.1.1 Symmetric Multiprocessing (SMP)

While these are technically multiprocessor systems, from the user's point-of-view they have properties of both single and multiprocessor systems. SMP systems consist of several processors that are connected to a common address space. Programs executing in that address space can run using any of the attached processors. In the sense that the address space is common, several threads that share an address space can actually be executing on different processors. Thus, although it would appear that data exchange could be done using the same mechanism used for single processor systems, it will not work. The protection afforded by disabling interrupts or disabling scheduling will not prevent a task running on another processor from continuing its operation. Data protection must therefore be based on access restrictions rather than scheduling ones. Implementation of data access restrictions generally incurs more overhead than scheduling restrictions. Alternatively, the processes can set up data exchange areas as if they were completely external. Copying data from there to the actual task data areas would then be under control of functions that could be controlled by scheduling restrictions. This, however, also involves higher overhead than single process data exchange.

These configurations are available on high-end workstations. Because they depend on connection to a common memory system the processors must be physically located in the place. They are used solely for the purpose of improving execution speed, not for geographic distribution of computing activities. Typically two to eight processors are used in SMP configurations.

11.1.2 Buses

In computer technology, a *bus* is a set of parallel wires that are used to share electrical signals among a group of devices. By using a large number of wires the data transmission rate can be very high because many bits are transmitted simultaneously. Using many wires in parallel, however, limits buses to relatively short distances because of the expense of running multiple wires over long distances and the difficulties of maintaining signal integrity.

There are two types of buses of interest for control systems: (1) computer component interconnect buses and (2) external device buses.

The first type of bus usually uses a large number of wires (50 to 100) with very high data rates to interconnect the circuit boards that make up a computer system. Common buses of this type are the ISA bus (the PC-AT bus used as a standard until the mid-1990s), the PCI bus (a faster replacement for the ISA), Multibus I and II and VME bus (used commonly in industrial systems), and several others. The boards that connect to such buses include processor boards, network interfaces, analog input/output interfaces, video boards, and disk controllers. Because of the high data rates and large number of wires, these buses are normally limited to very short distances. Thus, as with SMP, these configurations are useful for increasing the available computing power at a single site but not for geographically distributing computing resources.

These buses come in two flavors: single master and multimaster. Single master buses have no means for adjudicating simultaneous requests for access to the bus while multimaster buses have circuits to recognize and control such situations. Multimaster buses are most applicable to multiprocessor control systems since independent bus access is usually needed by all of the processors. Because all of the boards are in direct electrical contact, only one can actually write to the bus at any time.

The most common communication path between processor boards is to use shared memory. The bus defines a global address space that all of the boards can access. This space may or may not be directly accessible from programs running on each of the processor boards. If it is, then standard language constructs such as pointers can be used for data exchange. However, the same problems arise for this architecture as for symmetric multiprocessor systems in that data integrity protection must recognize that tasks on separate processors proceed independently. Although the extra data integrity protection adds overhead, the transfer rate of information across a bus is very fast (almost as fast as with SMP systems) so there is little delay associated with interprocess communication.

The second form of bus is designed for external use. It typically has fewer wires than an internal IO bus, and operates at lower speed. The two most common buses of this type are SCSI (small computer system interface) and GPIB (general purpose instrument bus, also called IEEE-488 or HPIB). SCSI buses are used almost exclusively for connecting disk drives and similar peripherals to computers. GPIB is normally used with a single computer and attached laboratory instruments (such as oscilloscopes). Neither is normally used for distributing control computation.

11.1.3 Networks

A broad definition of a network is a shared, serial communication channel. Networks are generally used for relatively long distance communications so the use of a serial channel is economically essential.[1] A serial channel transmits one bit at a time, as compared to a bus which can transmit as many bits simultaneously as there are data wires. The information rate on a network depends on the technology used and varies over an extremely wide range. At the low end, networks based on standard serial protocols RS232 and RS422 have data rates ranging from a few thousand bits per second to around 100,000 bits per second. Direct connect ranges can be up to

---

[1] There are only a few exceptions to this rule, such as the PLIP protocol which allows point-to-point TCP/IP connections via parallel printer ports.

a few thousand feet, but can be extended with repeaters. At the other end of the spectrum, fiber optic cables and microwave links can transmit data at gigabit per second rates over extended distances.

What probably most uniquely characterizes networks is *packets*. Information that a computer at a network node wishes to transmit to another computer (or computers) is bundled into one or more packets before being put onto the network for transmission. The packets serve the function of envelopes in ordinary mail. The "outside" has the information that the network management systems need to make sure the packet is sent to the right place, and the "inside" contains the data being transmitted. Once the packet arrives, it is decoded and the information in it is sent to whatever application program is expecting it. This is analogous to opening the envelope and delivering the contents. The actual nature of the information is not relevant to the network management layer. It could be numbers, code for a program, multimedia data, or any other information.

The International Organization for Standardization (ISO) defined an abstract model of a network that became the Open Systems Interconnection (OSI) reference model. It is a layered model that separates network functionality so that maximum portability of network hardware and software can be maintained. The model has seven layers:

- Application
- Presentation
- Session
- Transport
- Network
- Data-Link
- Physical

Information (or at the physical layer, a set of signals) is passed from one layer to another according to established protocols (or rules). As long as the protocols are well designed, creators of network hardware and software and applications retain the maximum flexibility and potential for creativity within layers. As has been discussed earlier, agreeing to operate according to certain rules or conventions does limit the amount of global optimization that can be done. More details on the role of each layer can be found in any of a number of texts on networking — much of this discussion and what follows is from Jamsa and Cope[16] and Bonner[17].

A property that distinguishes network architectures is how they handle the issue of allocating network bandwidth to all of the computers (nodes) on the network. There are two fundamentally different philosophies of doing this — rotating access permission and collision detection. Rotating access gives each node on the network permission in turn to write to the network. This is equivalent to a single master bus, with each node getting a turn at being master. Token ring systems are of this type, where the token is the permission to put information on the network. In systems based on collision detection, when a node wants to write to the network it first listens to see if there is any activity. If there is none, it proceeds to write its information to the network. There is a chance that more than one node will do this at the same time (i.e., within the time window between listening and starting to write). To detect this collision situation, each of the nodes listens to the network as it is writing to see if any information other than what it is writing is there. If it is, there has been a collision, and therefore the information has become corrupted. All of the affected nodes then stop transmitting, and each waits a random time interval before trying again.

The advantage of token ring architectures is order and predictability, whereas the advantage of collision detection is easy management and configuration. Although the predictability of performance for token ring architectures is cited as a reason to prefer them in a control system environment, both types of architectures have been used successfully.

The currently most popular network architecture is based on Ethernet lower layers (which use collision detection) and TCP/IP protocols for the upper layers. In its most popular forms, it uses either coaxial cable or twisted-pair wiring and achieves maximum data rates of 100 Mb/sec. The TCP/IP protocol is the basis for the Internet. TCP/IP will be the basis for the distributed control implementation given in this chapter and in chapter 12 so further relevant details will be discussed there.

Another network architecture that is becoming important in control systems is the instrument bus, also called the fieldbus. This is a relatively slow network, designed for connection of instruments and actuators to control computers. Because many of these elements are simple (a switch for example) keeping the per node cost very low is an important design goal for instrument networks. Some of the currently available network architectures in this group are CAN bus, Device Net, Profibus, and Interbus-S.

11.1.4 Point-to-Point Connections

Point-to-point connections use a unique physical communication channel for each computer-to-computer path. This has the advantage that each channel is wholly devoted to a single connection so that loads can be computed in advance and are not subject to interference with other communication paths. The disadvantage is that dedicated channels can only be used if the number of independent computers is small; otherwise, the number of channels needed rapidly gets out of hand. The number of adapters or ports needed to allow direct communication between all processors (each adapter is the connection point at one end of a channel) is $n*(n-1)$ where n is the number of computers to be interconnected.

The most common physical connection for point-to-point systems is RS232 or RS422 serial links. These are very similar, and are indistinguishable from the software. The difference is that RS232 uses a common ground topology while RS422 uses differential signals. RS422 has better noise immunity and can be run longer distances without a repeater, but requires more wires than RS232. These connection standards are very low cost and very well understood. From the network layer point of view, however, the standards do not include the upper layers so protocols for information transfer are often implemented *ad hoc*.

11.2 TCP/IP Networking

TCP/IP (Transport Control Protocol/Internet Protocol) networking has been chosen as the sample network technology because it is widely available on the types of PCs and workstations used for prototyping, its programming is well known, and programming tools are readily available for use with C and C++. The name TCP/IP is a bit confusing because it implies only the presence of the two named protocols. This is not the case, there are several other protocols that also exist within TCP/IP, but the name is commonly used to cover them all. TCP/IP is most commonly implemented locally using Ethernet as its physical layer, which uses collision detection (carrier sense multiple access with collision detection, CSMA/CD).

TCP/IP is built on a layered model similar to the ISO/OSI model, but with only four layers:

- Application

- Transport

- Network

- Link

For use in the real-time control environment, we will define an application layer. This custom application layer will then use standard protocols for all of the lower layers to preserve portability. The application layer will have to handle the two means of intertask communication defined in the previous chapter: messages and the shared variable database.

11.2.1 The Physical Context

In order to assure reliable delivery of information and predictable performance, it is assumed that the control system is implemented on a *private* TCP/IP network—that is, a network that has *no* direct, physical connections to external networks. Any communication that needs to be done with outside networks will be treated as a completely separate task using physically separate access to those networks. This is in contrast to most commercial implementations of TCP/IP which, ultimately, are connected to the entire world through the Internet.

11.2.2 Interconnection Protocols

Like all networks, the underlying information transport on TCP/IP networks is bundled into packets. This occurs at the network layer. At the transport layer, however, two protocols are available, one which is also packet oriented (User Datagram Protocol, UDP) and the other which is stream oriented (TCP). Before discussing them, however, the concepts of an unreliable channel and a reliable protocol need to be introduced.

TCP/IP networking was designed to work over highly unpredictable and unreliable communication channels. A packet targeted for a computer several thousand miles away will have to pass through many interconnects on the way in an environment of unpredictable load. A complete message could consist of many packets. Given all of this, it is possible that a packet might never get to its destination, or that it might get there but be garbled.

With this in mind, a reliable protocol is defined as one which can operate reliably in spite of the vagaries of an unreliable environment. To do this, error detection and correction information is added to the information being sent, and a handshake procedure is set up so that the sender can be assured that the information was received correctly. If a packet doesn't arrive, or arrives with too many errors to correct, it can be resent. The handshake procedure thus guarantees that if all else fails (i.e., if the error correction doesn't work or the packet is lost) the information is transmitted again. A reliable protocol also assumes that for cases of information contained in more than one packet, the order of arrival of the packets is not predictable either, so they will have to be put back in order after receipt.

The TCP protocol is defined as a reliable protocol while the UDP protocol is defined as unreliable. On the other hand, the control system network environment has a reliable physical layer so a protocol designed to operate in an unreliable environment is not necessary and will, in fact, add considerable overhead. In the case of control systems, the UDP protocol, the unreliable one, will be used. It is better suited structurally to the control problem as defined here and is much simpler to use. Coupled with a reliable link layer, it should be appropriate for many control projects. TCP and UDP will be described briefly, then the use of the UDP protocol for real time control will be explained in detail.

11.2.3 TCP and UDP

Programmers are used to writing code that communicates with external devices in a stream of characters (bytes). This format was formalized in very early versions of UNIX and has been widely adopted in other environments. The original development of TCP/IP was done on UNIX, so there was strong motivation to preserve the stream model. TCP was designed to fit this need. It establishes a pseudo (dare we say virtual?) connection between a receiver and sender. It is not a real connection because the underlying network and link layers are packetized. However, the software can be written as if there were a connection. Once a connection is established data can be sent whenever desired, just as if the data were going to a disk, printer, or a serial port. Because the receiving program is expecting a stream of data, TCP must have the means of correcting errors, putting packets in correct order, and requesting retransmission of missing packets. Thus, it is a reliable protocol which will either deliver correct data or indicate via an error return that it cannot.

UDP is connectionless. Data is sent in packets, there is no error correction or detection above the network layer, and there is no handshake. A packet is launched toward its target without any prior knowledge on the part of the target. If it doesn't arrive, the target has no way of knowing it was being sent and so cannot request retransmission. Although this sounds dire, UDP is a very useful protocol and does not need elaborate error correction and handshake if the underlying layers are reliable. In the controls case, the physical connection is reliable and all of the computers are connected on the same local link. Furthermore, the network traffic is very predictable so overload (and therefore a high potential for lost packets) is very unlikely. Finally, most of the data being sent are updates of variable values — if one update cycle is missed it will be updated on the next cycle. What this means is that if a higher level of reliability is needed for a particular communication, it can be implemented at the application layer. If a message is sent, for example, the state structure can be built so as to not proceed until a response is received. The packet structure of UDP is also particularly convenient since the message information and the database update information both fit naturally into a packet format.

11.2.4 Client/Server Architecture

Both TCP and UDP are built on a client/server model. In any given interaction, the server side waits for data to arrive. The data is often a request for information of some sort, which the server returns to the client. The model is therefore that the client begins all interactions. The server is passive until a request arrives. Every connection is set up with one side as the server and the other as the client. Each connection is unique — a single program can have some server roles and some client roles.

The client/server configuration is almost universally used for business computing applications and so is the standard for most networking protocols. It is not particularly well-suited to control system communication but can be made to work satisfactorily.

11.3 Implementation of UDP

Sample programs for this and subsequent sections are in the Net_Test directory of the grp_cpp.zip archive.

11.3.1 Sockets

Programming for network access is based on an abstraction known as a socket. The socket establishes a connection between the program and the system software needed to send and receive information across the network. The original design of

Section 11.3. Implementation of UDP

the network software for TCP/IP was intended to make use of the network look a lot like using files or other conventional I/O devices. Networks have some unique needs that don't quite fit into the file data-stream model, so the socket was invented to fill that role. A program can establish as many sockets as it needs. Each socket can be used to send and/or receive information.

Fewer sockets are usually required when using UDP than with TCP. Since TCP establishes a formal connection from one computer to another, it will require separate sockets for each computer with which it connects. UDP, on the other hand, is not connection oriented. The data packet rather than a continuing stream is its base unit of data exchange so the same socket can serve several purposes.

As noted, the original TCP/IP programming model was done in a Unix environment. It was subsequently adapted to the Windows environment where it is called the Winsock programming model. The examples here conform to the Winsock model.

11.3.2 Setting Up for Network Data Exchange

The setup for the server and client sides are almost the same. The example will be shown first for the server side. Then the few small changes for the client side will be shown. The procedure is to first initialize the network software so it can be used (the samples that follow are from `server1.cpp`):

```
#include <winsock.h>// Winsock header file

#define WINSOCK_VERSION 0x0101 // Program requires Winsock version 1.1

WSADATA wsaData;

...
if (WSAStartup(WINSOCK_VERSION, &wsaData))
{
cout << "Could not load Windows Sockets DLL." <<
PROG_NAME << " " << "\n";
exit(1);
}
```

The startup call doesn't have much substance—either it works or it doesn't. Once the network software has been initialized, sockets can be established. The function call defines a socket based on the Internet protocol family (PF_INET), a datagram (i.e, packet oriented) type of socket (SOCK_DGRAM), and one that uses UDP (IPPROTO_UDP). The parameters are defined in the file `winsock.h` which is supplied by the compiler manufacturer.

Note that these examples will only show the argument selections needed to implement the distributed control system. Refer to any book on Internet programming for detailed descriptions of the options available in more general circumstances.

```
SOCKET nSocket;   // Socket handle used by this program
...
nSocket = socket(PF_INET, SOCK_DGRAM, IPPROTO_UDP);
```

```
if (nSocket == INVALID_SOCKET)
{
cout << "Invalid socket!!" << PROG_NAME << " "
  << "\n";
exit(2);
}
```

The next step is to prepare the socket to receive information. A socket address structure is defined as in the following example. The address family defines for the socket the format that will be used for address. This is present so that the socket programming model could be used for a number of different network protocols. At the moment, only the Internet protocol is used, so the value AF_INET is the only value available.

The next value, the port, plays two roles in TCP/IP programming. Technically, it is just an identifier so that when information comes in to the network interface on a computer the system-level software can figure out where to send it. The Internet address (there will be more detail on this later) determines which computer the information should go to. The port determines which socket within the computer gets the information. That is the role that the port number plays here. The number chosen is arbitrary, but must be made known to the program sending the information so it goes to the right place.

For the application being discussed here, distributed control, the application layer is being uniquely designed. Many Internet programming projects, however, involve using one of the many standard application layer protocols such as File Transfer Protocol (ftp), mail, telnet, and World Wide Web (http). In order to communicate across the Internet to, for example, start a file transfer using ftp, the originating computer (the client) must be able to get a message to the ftp server program on the computer from (or to) which it wishes to transfer a file. It does this by using the standard port number for ftp. A set of these port numbers is published so that the standard protocols can be used on an anywhere-to-anywhere basis (in the literature, these are called the well-known ports).

The final value is used to restrict the Internet addresses from which this socket will accept messages. In this case, no restriction is needed, so the parameter is set to a default that will accept information from any address (INADDR_ANY).

```
SOCKADDR_IN sockAddr,RecvSockAddr;
...
// Define the socket address
sockAddr.sin_family = AF_INET; // Internet address family
sockAddr.sin_port = 1234;
sockAddr.sin_addr.s_addr = INADDR_ANY;
```

With the call to bind() the socket is set up to receive information. Its port is defined and the Internet addresses from which it will accept information is also defined. As in almost all subsequent network-related function calls, the socket number is the first argument.

```
nBind = bind(nSocket,(LPSOCKADDR) &sockAddr,sizeof(sockAddr));
if(nBind == SOCKET_ERROR)
```

Section 11.3. Implementation of UDP

```
{
cout << "Can't Bind, nBind= " << nBind << "\n";
exit(3);
}
```

11.3.3 Nonblocking Network Calls

In their default versions, function calls that require network activity are blocking. That is, they wait for the network activity to complete before they return. For example, a call to receive information from the network will not return until the information has actually been received. This is indefinite since the information might never be sent. In this sense, the socket function calls parallel the standard C/C++ input functions `scanf()` and `cin`.

Unlike the other input functions, the socket functions can be set to behave in a nonblocking fashion. If the requested action cannot be completed, the function will return with an error indicating that situation. In addition, it is possible to test in advance whether, for example, information is available. Combining these, the socket calls can be done such that the nonblocking restriction for all components of real-time software can be honored.

```
unsigned long NoBlock = TRUE,BytesToRead = 0;
int RetVal = ioctlsocket(nSocket, FIONBIO, &NoBlock);
if(RetVal == SOCKET_ERROR)
    {
    cout << " Error in setting socket to non-blocking\n";
    exit(4);
    }
```

This call sets the indicated socket to nonblocking status. Thus, even if for some reason a socket function were to be called at a time when it cannot complete its operation, it would return rather than block.

11.3.4 Receiving Information

Receiving information requires two steps, the first of which is checking to see if information is available. If new data has arrived, then it must be retrieved. For UDP, the protocol being used for control programs, the unit of information is a datagram packet. When a packet has arrived, the availability test will return positive. The test is:

```
RetVal = ioctlsocket(nSocket, FIONREAD, &BytesToRead);
if(RetVal == SOCKET_ERROR)
    {
    cout << "Error waiting for message\n";
    exit(5);
    }
}
```

The variable `BytesToRead` will return with a value greater than zero if a packet has arrived. The server program, `server1`, waits in a loop until this happens. In a control program, the loop would be provided by the state scan mechanism.

When information is ready, it is read by:

```
nCharRecv = recvfrom(nSocket,(char *)MessageBuffer,
    MESSAGE_BUFFER_SIZE,
    RECV_FLAGS,(LPSOCKADDR)&RecvSockAddr,
    &lenRecvSockAddr);
```

The first argument identifies the socket. The next two provide a buffer to copy the received information into.

The last two arguments get the Internet address from which the information was sent. In normal client/server relationships, the server needs to know this information so it can respond, usually by sending back the requested information. For control programs, however, the client sends information to the server, for example, updated values of shared database variables. No response is needed, however, unless some form of error checking were to be added. Thus, control programs do not use the full client/server architecture.

In this case, the sender's address is printed for diagnostic purposes:

```
AddrCon.IPAddr = RecvSockAddr.sin_addr.s_addr;
cout << "Sender Address " << (int)AddrCon.bbb.b1
    << " " << (int)AddrCon.bbb.b2 << " "
    << (int)AddrCon.bbb.b3 << " " << (int)AddrCon.bbb.b4
    << " " << "\n";
```

The sender's Internet address is a 32-bit unsigned integer in its original form. It is decoded into the familiar dotted-decimal notation (e.g., 136.152.66.28) by the union

```
union  // To get dotted-decimal addresses
       //from 32-bit addresses
  {
  u_long IPAddr;
  struct
    {
    u_char b1,b2,b3,b4;
    }bbb;
  }AddrCon;
```

The casts to `int` in the `cout()` statement are used because `cout()` will interpret the `u_char` (unsigned char) data type as an ASCII character unless the cast is applied.

11.3.5 Client-Side Setup

The setup for the client program is almost the same as for the server program (the sample program is `client1.cpp`). Since it initiates a data transfer, it must know what the Internet address and port number are for the recipient. Also, it sends rather than receives data.

The Internet address of the recipient is specified in dotted decimal format as a character string:

```
#define SERVER_ADDRESS "136.152.66.28"
```

This format must be converted to the internal format to be used:

```
ServAddr = inet_addr(SERVER_ADDRESS);
```

The `bind()` function call appears to be important for the client program, although in theory it can function without it because control clients' sockets only send; they do not receive anything. As noted, a program can have several sockets, some of which are acting as servers, others as clients. These sample programs only have one socket each.

When setting data for binding, it is wise to use a different port number for each process. Although this is not necessary in actual use, it is very useful in testing where all of the processes can be run on the same computer and will thus all have the same Internet address. They can then only be identified by their port numbers.

To send information, the address information for the recipient must be specified:

```
SendSockAddr.sin_family = AF_INET;   // Internet address family
SendSockAddr.sin_port = 1234;
SendSockAddr.sin_addr.s_addr = ServAddr;
```

where `ServAddr` is the address resulting from the conversion. The port must match the port used by the recipient socket in its `bind()` call.

Sending data is done with the `sendto()` function:

```
nCharSent = sendto(nSocket,(char *)MessageBuffer,nbuf,
    SEND_FLAGS,(LPSOCKADDR)&SendSockAddr,sizeof(SendSockAddr));
```

`Sendto()` is naturally nonblocking because all it does is give the data to the lower level network functions where it is buffered for sending. It thus does not have to wait for any network activity and returns as soon as that operation is complete.

11.4 The Application Layer

With all of this network activity, nothing has yet been said about the nature of the information being interchanged. That is the domain of the application layer.

A very simple application layer is defined for this sample program set. It includes the ability to send several standard data types as single (scalar) values or as arrays. A packet is made up of as many sets of these data as desired (only a few of the standard data type items are actually implemented).

11.4.1 Data Coding

Before looking at the process of assembling a packet, however, an ugly reality must be faced: each of the processes making up a control program could be on a different type of computer. Different computers code numbers differently so information for proper decoding must be included in the packet. Even numbers as simple as integers do not have a uniform coding format. While almost all computers in common use now use 2's complement notation, some are big endian and others are little endian, referring to the order in which the bytes constituting an integer are stored (highest order byte first or lowest order byte first).

About the only format with any hope of universality is a single byte coded to be read as an unsigned integer (data type unsigned char). The first byte of the packet, coded as unsigned char, is used to identify the format for all of the following data. The only format thus far defined is for Intel x86 computers using the Borland C++ compiler. It has been assigned a code of 0. No actual conversion routines are necessary in this circumstance, but as long as the data format identifier is present, they can be added.

11.4.2 Building the Packet

Information is converted from the native data type to a stream of bytes by use of unions that define the same memory space in more than one way. The packet is built in the client program (client1.cpp) for sending to the server. For example, to convert a short (16-bit integer) to a byte stream the following union is used:

```
union pack_short
    {
    short s;
    unsigned char c[2];
    };
```

To do the conversion, a short integer variable is assigned to s. The bytes in the array c can then be copied to the buffer for the packet being assembled. Here is the function that inserts a short integer into the buffer:

```
int InsertShortValue(unsigned char *packet_array,int npack,int value)
    {
    union pack_short ps;
    int i = npack;

    if((int)((short)value) != value)
        {
        cout << "<InsertShortValue> Truncation error\n";
        exit(6);
        }
    packet_array[i++] = 1;   // Code for single short value
    ps.s = (short)value;
    packet_array[i++] = ps.c[0];
    packet_array[i++] = ps.c[1];
    return(i);
    }
```

Before the actual data is inserted into the buffer, a code (unsigned char) is entered to identify the type of data that is being stored. The set of codes defined is (although insertion routines are only provided for a few of these):

Section 11.4. The Application Layer

```
Block codes:
  Single values (value follows directly)
    0  unsigned char
    1  short
    2  long
    3  float
    4  double
    5  unsigned short
    6  unsigned long
  Arrays (number of values is next)
    7  unsigned chars
    8  shorts
    9  longs
   10  floats
   11  doubles
   12  unsigned shorts
   13  unsigned longs
  Structures
   10  standard message (...)
```

For scalar values, the actual value follows directly after the code. For arrays, a short integer value follows to specify the number of values in the (single dimensional) array. The function to store an array of doubles is:

```
int InsertDoubleArray(unsigned char *packet_array,int npack,
        double *array,int nval)
    {
    union pack_double pd;
    union pack_short ps;
    int i = npack;

    packet_array[i++] = 11;  // Code for array of doubles

    ps.s = (short)nval;     // Enter number of values in array
    packet_array[i++] = ps.c[0];
    packet_array[i++] = ps.c[1];

    for(int k = 0; k < nval; k++)  // Enter the array values
        {
        pd.d = array[k];
        for(int j = 0; j < 8; j++)packet_array[i++] = pd.c[j];
        }
    return(i);
    }
```

These functions are called repeatedly to build the packet for transmission. Here is the loop from `client1.cpp` in which three packets are built and then transmitted. Each packet contains a short integer, a double, and an array of two doubles. In each case, the values are manipulated by the loop index so that the three packets contain different information.

```
...
for(int i = 0; i < 3; i++)
  {
  // Send several packets
```

```
        // Initial value in buffer indicates the type of data coding used
    nbuf = 0; // Count number of characters in the buffer
    MessageBuffer[nbuf++] = 0; // Code indicating data is in Intelx86
        // (Borland) format

    nbuf = InsertShortValue(MessageBuffer,nbuf,14*(i + 1));
    nbuf = InsertDoubleValue(MessageBuffer,nbuf,234.56 * (i + 1));
    vv[0] *= (i + 1);
    vv[1] *= (i + 1);
    nbuf = InsertDoubleArray(MessageBuffer,nbuf,vv,2);
    cout << "nbuf " << nbuf << " ";

    nCharSent = sendto(nSocket,(char *)MessageBuffer,nbuf,
        SEND_FLAGS,(LPSOCKADDR)&SendSockAddr,sizeof(SendSockAddr));
    cout << "nCharSent " << nCharSent << "\n";
    }
```

This generates the following screen image when executed (in Windows NT as a console mode application):

```
ServAddr 474126472
 nBind= 0
nbuf 32 nCharSent 32
nbuf 32 nCharSent 32
nbuf 32 nCharSent 32
Hit any key to exit.
```

The `ServAddr` and `nBind` values are present for debugging purposes.

11.4.3 Parsing a Packet

When a packet arrives at the server the inverse process must be followed. The packet has to be taken apart (parsed) into its component pieces. This process is very similar to building a packet. It uses the same unions to overlap sequences of bytes with standard data types. For example, here is the function that extracts an array of doubles from a packet:

```
int ExtractDoubleArray(unsigned char *packet_array,int next,
        double *pvalue,int *pnval)
    {
    int i = next;
    int nval,j;
    double v;

    i = ExtractShortValue(packet_array,i,&nval);  // Get number of values
    *pnval = nval;
    for(j = 0; j < nval; j++)
        {
        i = ExtractDoubleValue(packet_array,i,&v);
        pvalue[j] = v;   // store array value
        }
    return(i);
    }
```

Section 11.4. The Application Layer

The code to parse a packet is a loop that looks for the item type, then uses the appropriate decoding method. Here is the parsing function from `server1.cpp`:

```
int ParseMessage(unsigned char *Buf,int nCharRecv)
    {
    // Break out component parts of the message and print results
    // Return 0 for OK, -1 for error
    int nused = 0;  // Number of characters used in decoding buffer

    if(Buf[nused++] != 0)
        {
        // Incorrect data format (Only Intelx86 recognized here)
        cout << "<ParseMessage> Incorrect data format, 1st char = "
        << (int)Buf[0] << "\n";
        return(-1);
        }
    while(nused < nCharRecv)
        {
        // Continue this loop until the message is entirely consumed
        int BlockType = Buf[nused++];

        switch(BlockType)
            {
            case 0:  // unsigned char
            cout << "Unsigned char = " << (int)Buf[nused++] << " ";
            break;

            case 1:  // short
                {
                int value;

                nused = ExtractShortValue(Buf,nused,&value);
                cout << "Short = " << value << " ";
                break;
                }

            case 4:  // double
                {

                double value;

                nused = ExtractDoubleValue(Buf,nused,&value);
                cout << "Double = " << value << " ";
                break;
                }

            case 11:  // array of doubles
                {
                int n;
                double v[10];

                nused = ExtractDoubleArray(Buf,nused,v,&n);
                cout << "Array of Doubles ";
                for(int i = 0;i < n; i++)cout << v[i] << " ";
                break;
                }
```

```
            default:
                // No other type implemented yet!
                cout << "\nUnknown Block Type\n";
                return(-1);
            }
    }
    cout << "\nEnd of Packet\n";
    return(0);
}
```

Note that not all of the item types are handled in this, just a representative sample. Also, only the Intel x86 data format is handled; any other format triggers an error.

The screen output for the three packets sent by `client1.cpp` is:

```
nBind= 0
Hit any key to exit.
nCharRecv 32
Sender Address 136 152 66 28
Short = 14 Double = 234.56 Array of Doubles 1.3 0.00521
End of Packet
nCharRecv 32
Sender Address 136 152 66 28
Short = 28 Double = 469.12 Array of Doubles 2.6 0.01042
End of Packet
nCharRecv 32
Sender Address 136 152 66 28
Short = 42 Double = 703.68 Array of Doubles 7.8 0.03126
End of Packet
```

This completes the development of a simple client/server pair utilizing the UDP. It implements a simple application layer, but one that will serve as a base for the control system protocol that will be developed in the next chapter.

Chapter 12

DISTRIBUTED CONTROL II: A MECHATRONICS CONTROL APPLICATION LAYER

The material in chapter 11 established a framework for network communication of information of the type needed for control of mechatronic systems. It did not, however, actually establish an application layer that could be integrated into control system software. This chapter does that by building on the UDP software and the simple application layer that was defined to demonstrate its application.

Development of the application layer is done in two stages. In the first, the application layer protocols are defined and a test program is constructed to implement the test protocols. In the second stage, the application layer is integrated with the existing control software.

12.1 Control System Application Protocol

The software described in this section sets up the application protocol and uses small test programs to verify that data of the proper format is actually transmitted and received.[1] The functions that do most of the work, and that will form the basis for the network interface in the actual control software, are in the file `net_udp.cpp`. The test programs that exercise these functions are in the files `test0.cpp` and `test1.cpp`.

The UDP block used has the same basic format defined in the previous chapter. The major difference is that only two primary data block types are defined, one for messages and one for shared data arrays. Additional block types are defined to deal with the start-up synchronization problem; this will be described later.

---

[1] The sample code for this section is found in the `grp_cpp` archive, directory `net_UDP`.

A UDP packet is made up of any number of data blocks. Each block is identified as to type and then follows the rules for storing data of that type. As before, the UDP packet begins with a byte indicating the format protocol used for coding the data. The existing sample software only recognizes Intel x86 format, type 0. A data packet thus looks like,

```
Data coding (1 byte)
Block type (1 byte)
<Message block or Shared Data block>
Data coding (1 byte)
Block type (1 byte)
<Message block or Shared Data block>
...
```

The data format within a block is best described by the code used to decode blocks. A message block in the packet is decoded (parsed) by the following code (in the file net_udp.cpp):

```
case 1:   // Message block
    {
    int ToProc,ToTask,BoxNo,FromProc,FromTask;
    int MsgFlag,RetRecpt,DelMode;
    double MsgValue;

    nused = ExtractShortValue(Buf,nused,&ToProc);
    nused = ExtractShortValue(Buf,nused,&ToTask);
    nused = ExtractShortValue(Buf,nused,&BoxNo);
    nused = ExtractShortValue(Buf,nused,&FromProc);
    nused = ExtractShortValue(Buf,nused,&FromTask);
    nused = ExtractShortValue(Buf,nused,&MsgFlag);
    nused = ExtractDoubleValue(Buf,nused,&MsgValue);
    nused = ExtractShortValue(Buf,nused,&RetRecpt);
    nused = ExtractShortValue(Buf,nused,&DelMode);

    StoreMessage(ToProc,ToTask,BoxNo,FromProc,
                 FromTask,MsgFlag,MsgValue,RetRecpt,DelMode);
    break;
    }
```

The StoreMessage() function would normally be in basetask.cpp and extends the message model introduced earlier to include receipt of messages from external sources. For this test code however, a dummy version is supplied in the test programs, test0.cpp and test1.cpp. It prints the received values so they can be checked:

```
// Dummy routines for the control program interface routines
void StoreMessage(int ToProc,int ToTask,int BoxNo,int FromProc,
    int FromTask,int MsgFlag,double MsgValue,int RetRecpt,int DelMode)
    {
    cout << "***Message received, Process #" << ThisProcess << "\n";
    cout << "ToProc " << ToProc << " ToTask " << ToTask << " BoxNo "
        << BoxNo << "\n";
```

Section 12.1. Control System Application Protocol

```
        cout << "FromProc " << FromProc << " FromTask " << FromTask << "\n";
        cout << "MsgFlag " << MsgFlag << " MsgValue " << MsgValue
            << " RetRecpt " << RetRecpt << " DelMode " << DelMode << "\n";
    }
```

The protocol for receiving a shared data array is:

```
            case 2:    // Shared value array
                {
                int n,ToProc,ToTask;

                nused = ExtractShortValue(Buf,nused,&ToProc);
                nused = ExtractShortValue(Buf,nused,&ToTask);
                nused = ExtractDoubleArray(Buf,nused,&v,&n,&vSize);

                StoreSharedData(ToProc,ToTask,n,v);
                break;
                }
```

and the associated dummy function for printing that data is:

```
void StoreSharedData(int ToProc,int ToTask,int n,double *v)
    {
    cout << "***Shared Data received, Process #" << ThisProcess << "\n";
    cout << "ToProc " << ToProc << " ToTask " << ToTask << "\n";
    for(int i = 0; i < n; i++)cout << v[i] << " ";
    cout << "\n";
    }
```

The complementary operations, used to assemble the UDP packet, are as follows (also from **net_udp.cpp**). They take the information about the message or shared data block and insert it into a memory buffer (`PacketBuffer`) used to store a UDP packet prior to actual transmission.

For assembling a message block:

```
int InsertMessage(int ToProc,int ToTask,int BoxNo,int FromProc,
                  int FromTask,int MsgFlag,double MsgValue,
                  int RetRecpt,int DelMode)
    {
    // Returns error: 1 for error, 0 for OK
    // Add a message to the appropriate process packet

    if(ToProc == ThisProcess)
        {
        cout << "<InsertMessage> Messages to this process\n"
            << "Should have been handled locally not in network\n";
        return(1);
        }
    int next = PacketNext[ToProc];
    if((next + 25) > PACKET_BUFFER_SIZE)
        {
        cout << "<InsertMessage> Packet overflow, process # "
            << ToProc << "\n";
        return(1);
        }
```

```
    unsigned char *pc = pPacketBuffer[ToProc];  // Pointer to buffer

    pc[next++] = 1;  // Code for message block
    next = InsertShortValue(pc,next,ToProc);
    next = InsertShortValue(pc,next,ToTask);
    next = InsertShortValue(pc,next,BoxNo);
    next = InsertShortValue(pc,next,FromProc);
    next = InsertShortValue(pc,next,FromTask);
    next = InsertShortValue(pc,next,MsgFlag);
    next = InsertDoubleValue(pc,next,MsgValue);
    next = InsertShortValue(pc,next,RetRecpt);
    next = InsertShortValue(pc,next,DelMode);

    PacketNext[ToProc] = next;  // Update packet index
    return(0);
}
```

The equivalent function for a shared data array is:

```
int InsertSharedArray(int ToProc,int ToTask,int nValues,double *Val)
{
// Returns error: 1 for error, 0 for OK

    if(ToProc == ThisProcess)
        {
        cout << "<InsertSharedArray> Data for this process\n"
            << "Should have been handled locally not in network\n";
        return(1);
        }
    int next = PacketNext[ToProc];
    if((next + 7 + nValues * sizeof(double)) > PACKET_BUFFER_SIZE)
        {
        cout << "<InsertSharedArray> Packet overflow, process # "
            << ToProc << "\n";
        return(1);
        }

    unsigned char *pc = pPacketBuffer[ToProc];  // Pointer to buffer

    pc[next++] = 2;  // Code for shared data array
    next = InsertShortValue(pc,next,ToProc);
    next = InsertShortValue(pc,next,ToTask);
    next = InsertDoubleArray(pc,next,Val,nValues);

    PacketNext[ToProc] = next;  // Update packet index
    return(0);
}
```

Additional block types are defined for control rather than data transmission purposes. These three are used for the startup process to assure that no control software is actually run until all of the processes are online. Each of these blocks is 3 bytes long:

```
Data type
Block type
Process number
```

These three block types are:

```
Type Function
100 Check to see if a process is on line ("Are you there?")
101 Confirm that a process is on line ("I am here")
102 Start regular operation when all processes have been confirmed
```

The process byte for type 100 and 102 is the target process (the process to which the block is being sent) and is used to confirm correct delivery. The process byte for type 101 is the process number of the process that is identifying itself to the system as being on line.

12.2 Startup of Distributed Control Systems

The startup procedure must assure that all processes constituting the control system are online before allowing any of the control code to run. For each process, this is done after the network sockets have been established.

Establishing the sockets is done in a manner very similar to that in which it was done in the previous chapter. A simple map must be established for each process so it knows process numbers and corresponding Internet addresses. This is done at the beginning of the test0.cpp and test1.cpp files:

```
static int ThisProcess = 0;  // Change this for other file
static int ToProc=1;
static char *ProcessAddress[2] =
    {
    // The same address for both processes is for testing
    //   with both processes running on the same computer
    "10.0.2.15", //"128.32.142.37", "10.0.2.15",
    "10.0.2.15" //"128.32.142.37" "10.0.2.15"
    };
```

The socket initialization is in the InitUDP() function in the file net_udp.cpp:

```
if (WSAStartup(WINSOCK_VERSION, &wsaData))
    {
    cout << "<InitUDP> Could not load WinSock DLL.\n";
    ErrorExit(1);
    }

nSocket = socket(PF_INET, SOCK_DGRAM, IPPROTO_UDP);
if (nSocket == INVALID_SOCKET)
    {
    cout << "<InitUDP> Invalid socket!!\n";
    NetCleanUp();
    ErrorExit(2);
    }
// Define the socket address
sockAddr.sin_family = AF_INET;  // Internet address family
sockAddr.sin_port = u_short(BASE_PORT + ThisProc);
    // Define the port for this process
sockAddr.sin_addr.s_addr = INADDR_ANY;
```

```
// Bind the socket
nBind = bind(nSocket,(LPSOCKADDR) &sockAddr,sizeof(sockAddr));
if(nBind == SOCKET_ERROR)
    {
    cout << "Can't Bind, nBind= " << nBind << "\n";
    NetCleanUp();
    ErrorExit(3);
    }

// Set socket to non-blocking
unsigned long NoBlock = TRUE;
int RetVal = ioctlsocket(nSocket, FIONBIO, &NoBlock);
if(RetVal == SOCKET_ERROR)

    {
    cout << "<InitUDP> Error in setting socket to non-blocking\n";
    NetCleanUp();
    ErrorExit(4);
    }
```

Memory is then allocated for the packet buffers (in `ResetPacket()`) and a set of flags indicating which processes are connected (`ProcessConnected()`) are initialized to 0 for nonconnection, except for the flag for the `ThisProcess`.

Once the basic network connections have been established, the next step is to synchronize all of the processes. To do this, Process #0 is arbitrarily assigned as the master. It contacts all other processes and when all processes have responded, it sends out a signal to start control. The procedure, which is implemented in `ContactProcesses()`, performs the following steps:

- Send out "Are you there" signals (UDP block type 100) to all processes not yet connected. Do this periodically to avoid overloading the network.

- Watch for "Here I am" (type 101) return messages. When one arrives, mark the associated process as connected.

- Send the "Start" (type 102) signal to all processes when all processes have responded.

In the meantime, all processes other than #0 watch for "Are you there" signals, send out the "Here I am" responses, then wait for the "Start" signal to begin their control activities.

12.3 Testing the Application Protocol

Each of the two test programs (`test0.cpp` and `test1.cpp`) sends out samples of message blocks and shared data blocks. The dummy `StoreMessage()` and `StoreSharedData()` functions print the incoming messages to the screen for debugging. Sending the data takes place as follows:

```
// Send data to the other process
int ToTask=2, nValues = 3;
double Val[] = {1.2,4.3e-4,5.4e2};
RetVal = InsertSharedArray(ToProc,ToTask,nValues,Val);
cout << "Insert Shared RetVal " << RetVal << "\n";

ToTask=2;
int BoxNo=3,FromProc=ThisProcess,FromTask=2;
int MsgFlag=7,RetRecpt=0,DelMode=1;
double MsgValue=9.87;
RetVal = InsertMessage(ToProc,ToTask,BoxNo,FromProc,
FromTask,MsgFlag,MsgValue,RetRecpt,DelMode);
cout << "Insert Message RetVal " << RetVal << "\n";
RetVal = SendData(); // Put the data out on the network

cout << "Send Data RetVal " << RetVal << "\n";

cout << "Hit any key to exit.\n";
while(1)  // Infinite loop - user causes exit
    {
    CheckNet();  // Check for incoming data
    if(kbhit())break ;
    }
NetCleanUp();
}
```

The resulting output screen is:

xxxxxxxxxxxxxxxxxx

12.4 Using the Control Application Protocol

Minimal program modification is needed to run in a multiprocess environment. Programs that are designed for multiprocess operation can be run in a single process environment with no change at all except for the process map.[2] Programs written originally for a single process environment need to be modified to make assignments of tasks to processes.

The major distinction in implementation is that a completely separate program must be built for each process. To maintain maximum portability the programs that are compiled are nearly identical. All of the code is compiled for each process. The only difference is that each task is executed in only one process. This is accomplished in the section of the program where tasks are added to each of the lists for execution. This is done with the aid of information from the process map, which is in `tasks.hpp`:

```
#define PROCESS_INFO // So that basetask.cpp does not put in defaults

#ifdef PROCESS_INFO  // Bring in process information; otherwise allow defaults
    #include "proc_no.hpp"  // Defines the process number for this process
```

---

[2] Samples are in directory **heater4** of the grp_cpp archive.

```
    const int NumberOfProcesses = 2;  // Number of independent processes
    // Process map:
    #define PROCESS_A 0  // Define both as 0 for single process system
    #define PROCESS_B 1
    #define USE_UDP  // Only define this if UDP protocols are to be used
#endif

#ifdef USE_UDP
    #include "net_udp.hpp"
    char *ProcessAddress[] =
        {
        // Process network addresses
        "10.0.2.15", //"128.32.142.37",  "10.0.2.15",
        "10.0.2.15" //"128.32.142.37"  "10.0.2.15"
        };
#endif
```

All possible processes that could be used are established by using #define to name the set of processes, PROCESS_A and PROCESS_B in this case. At the same time, each named process is assigned to an actual process number. Assigning all of the processes to the same number gives a single process implementation, and so on. The type of network protocol used is also defined at this point. For UDP, the only network protocol currently implemented, the Internet addresses, for each actual process must also be defined (in dotted-decimal format).

One additional piece of information is needed, the process number. This is unique for each process program to be compiled, and is the only unique information needed. A separate directory is used to compile each of the programs, one for each process. The file proc_no.hpp is the only program code in the directory:

```
// Define the process number for this process
// File: proc_no.hpp
// Each process has a unique file with this name
//   to define its process number.

int ThisProcess = 0;
```

The task assignment uses this information plus the process map (heatctrl.cpp):

```
// All tasks must be added to the task list here.
// Only add tasks that are in this process
#ifdef ISR_SCHEDULING
if(ThisProcess == PROCESS_A)Preemptable->Append(PWM);
if(ThisProcess == PROCESS_A)Interrupt->Append(HeatCtrl);
#else // ISR_SCHEDULING
if(ThisProcess == PROCESS_A)Intermittent->Append(PWM);
if(ThisProcess == PROCESS_A)LowPriority->Append(HeatCtrl);
#endif // ISR_SCHEDULING
if(ThisProcess == PROCESS_A)LowPriority->Append(HeatSup);
if(ThisProcess == PROCESS_B)LowPriority->Append(OpInt);
if(ThisProcess == PROCESS_A)LowPriority->Append(DataLogger);
#ifdef SIMULATION
if(ThisProcess == PROCESS_A)LowPriority->Append(SimTask);
#endif //SIMULATION
```

In this case, the operator interface is assigned to PROCESS_B while all other tasks are assigned to PROCESS_A.

The low level code (in basetask.cpp) takes care of filling the buffer with information to be sent out, or distributing information from packets that come in. Someplace must be found to execute the code that actually polls the network for incoming packets and sends the current outgoing packets to the network. The heater sample program does that by putting the code in the main function (in heatctrl.cpp) inside the loop that calls the cooperative dispatcher. This puts the network access software at the lowest priority level, equal to the LowPriority tasks. The dispatch loop then looks like:

```
while (((TheTime = GetTimeNow()) <= EndTime)
         && (TheTime >= 0.0) && (!StopControlNow))
{
// Run LowPriority and Intermittent list
if (TaskDispatcher(LowPriority,Intermittent))
   StopControl("\n***Problem with background tasks***\n");
#ifdef USE_UDP
  if(TheTime > NextNetTime)
     {
     CheckNet();  // This is the lowest priority position --
     // Higher priority positions may be more appropriate
     if(!SendAll())StopControl("\n***Problem sending network data***\n");
     NextNetTime += DeltaNetTime;
     }
#endif
if(OpInt->GetGlobalData(OpInt,OPINT_STOP) > 0.0)
  {
  SendAll();  // Make sure final data is sent (it has the
     // STOP signal!)
  StopControl("\n\n***User terminated program with ctrl-x***\n\n");
     // Do it this way so all processes will stop, not just the one with
     // the operator interface
  }
}
```

Data is sent to the network, and the network interface is polled for data at a specified time interval. The interval is chosen as a compromise between keeping the data as timely as possible and avoiding overload on the network. Another possibility would be to put this code in a task which is scheduled to run in all tasks. It could then be placed at any desired priority level.

Note that the means of stopping the program has also had to change from the way it was done in the previous, single-process examples. In the previous case, whichever task controlled stopping made a call to StopControl(). This will not work in the multiprocess case because that will only stop the process in which the call is made. None of the other processes will know about the stop signal. Stopping is now controlled by a global variable, which can be monitored by all of the processes. Each one then stops itself when it detects the global stop signal. Note also that there is an additional SendAll() function call after the stop signal has been detected. It is there to make sure that the stop signal is actually sent from the process that orginated the stop signal. Otherwise, that process would stop and would never send the global variable out to the other processes.

Running this program is identical to running the single process version. The only difference is in performance—several computers are running simultaneously, but information interchange among processes is limited by network delays.

12.5 Compiling

The program structure is set up so that the source code for multiprocess implementation is nearly identical to that for single process implementation. This makes experimentation with the implementation very easy—a primary goal of this methodology.

When the two processes use the same computational environment (as they do in this example) the only changes are in the assignment of tasks to processes. This leads to programs that are not as memory efficient as they could be since all tasks are included in all processes, but are only executed in one process. Part of the class structure for each task is used, however, even if the task is not being executed. Some memory usage efficiency improvement could be achieved by not compiling code or data that will not be used, but this would add considerable complication to the program.

This example uses a directory structure in which all of the source code (application code and code from the common directory) resides in one directory. Each process uses a daughter directory one level below the source code directory for its project file (*.ide) and for the include file that sets the process number. This include file (which only has one line of code) is the only source code that differs for each process. A separate executable file, one for each process, is then compiled in each of these directories.

Chapter 13

JAVA FOR CONTROL SYSTEM SOFTWARE

Java is a computer language originally developed for embedded applications by Sun Microsystems, Inc. Before it ever saw the light of day as an embedded system language, however, it was cleverly adapted for use on the then emergent Internet and World Wide Web. The original Internet hypertext transfer protocol (HTTP) was strictly for information retrieval. The only purpose for the client (browser) software was to display information fed to it in Hypertext Markup Language (HTML) by the server. As the Web developed, however, the slow speed associated with transactions taking place across the Internet cried out for a means for the server to direct the client to do some computing, with the results made available back to the server. While some scripting languages were available for simpler computing needs, there was nothing available for more complex, client-based computing. The Sun Java development group (led by James Gosling) realized that they had a language that could fill that role.

Security was already a very high priority issue for the Internet. The interpretive structure of Java meant that security issues could be handled in the interpreter as long as the features of the language were such that blatantly dangerous operations (such as those with pointers) were eliminated. Another important property that was well supported by the planned structure of Java was the need for portability. Internet browsers could be run on any kind of client computer, running any available operating system. As long as each browser had an interpreter for its own environment, the server did not have to worry about the type of computer or operating system it was communicating with.

These properties, particularly the portability, fit very well with the goals for control system development discussed here. Indeed, ever since the Java language was introduced, there has been interest in using it for real-time applications. However, it has only been recently that doing so was even somewhat practical. This chapter briefly discusses the benefits and drawbacks of using Java, its use in real-time programming, and some of the particular advantages for the development of control system software.

13.1 The Java Language and API

Since its introduction, there have been four new versions of Java, with the most recent being version 1.4. In reality, Java is no longer just a language, but rather a runtime and large collection of APIs, each targeted at a different segment of a continually growing market[18]. From corporate databases to the Web, Java is already practically everywhere.

It has taken several years for Java to return to its roots, but potential applications in embedded systems are now greater than ever. Even as embedded processors become cheaper and more prevalent, there is an increased demand for interconnectivity among them. This is true for both consumer and industrial devices, but there are special implications for the future of embedded control. The use of embedded Java could potentially result in smarter, more distributed control networks, due in part to some of the following language features.

13.1.1 Networking

Java has been a network enabled language from the beginning. Even the earliest versions of the java.net package provided developers with an easy-to-use API that allowed basic network connectivity. This simple framework has now expanded dramatically as Java has grown in popularity for enterprise software development, with its heavy reliance on distribution and interacting objects. Still, the basic java.net package provides all the functionality that is required to implement simple inter-object communication through either datagrams (UDP) or streams (TCP). Since all the underlying socket manipulation is performed by the java.net classes and native code, the result is much cleaner communication code, as well as independence from notoriously platform-dependent socket APIs. This means that Java is still uniquely suited for use with future communication protocols that embedded and control systems may require.

13.1.2 AWT/Swing

Another fundamentally unique aspect of Java is the Abstract Windowing Toolkit, or AWT. In the past, designing a GUI for a program was a platform-dependent process at best. Now it is possible, at least for the most part, to build a GUI once and deploy it on multiple platforms. The advantages of this can be dramatic, since interface design can be one of the most labor-intensive aspects of application design. When combined with Java's built-in downloadable applet support, the result is a powerful method of delivering client-side applications on demand.

13.1.3 Multithreading

In contrast to C and C++, the Java specification includes support for multithreading as part of the language itself. In a way, this has brought concurrent programming to the masses, since the Thread API is reasonably easy to work with. Java also manages to hide most of the most complex aspects of multithreading from the user.

For example, to ensure thread safety of data, Java provides for critical sections of code using the `synchronized` keyword. The interpreter then manages all thread access to protected areas automatically.

The primary drawback to the Java threading model is that it really doesn't exist. Although the Java thread classes present a uniform interface to programmers, the language specification says nothing about how threads are implemented in the virtual machine itself. Depending on the processor, operating system, and even vendor, the variation between different virtual machines can be dramatic. One striking example of the variation comes from Sun itself: the 1.1 version of its Java runtime for Windows NT used native OS threads while the Solaris implementation utilized a green-threading scheme. Needless to say, a multithreaded Java program could have substantially different operation in these two environments[19]. To further complicate matters, Java thread priorities often do not correspond directly to those of the underlying OS. For example, the 10 levels of Java thread priority map onto only seven levels under Windows NT. It is generally the case that Java thread priorities other than `PRIORITY_HIGHEST`, `PRIORITY_NORMAL`, and `PRIORITY_LOWEST` are not guaranteed to be any different[20].

13.2 Preconditions for Real-Time Programming in Java

Despite all the effort being put into new real-time specifications, there are really only a few necessary conditions that must be met for a Java application to run in real time. Note that the slower execution speed of virtual machine-based software is not, in itself, a barrier to real-time execution. The key to real-time software is determinism, not speed.

13.2.1 Deterministic Garbage Collection

Although developers may find memory management in Java to be much less stressful than in other languages, there are associated problems which significantly impact control software design. In particular, the memory reclamation scheme used by most commercial virtual machines is not at all suitable for real-time control. Typically referred to as the "garbage collector," there is a continuously operating thread that locates objects without any remaining valid references and destroys them. Asynchronous operation of the garbage collection thread can lead to priority inversion situations, since this internal VM thread can block for an unspecified amount of time. Recently, there have been a number of proprietary and clean-room virtual machines developed by third-party vendors specifically to provide less blocking incremental garbage collection.

13.2.2 Memory and Hardware Access

In an effort to maintain platform independence, the Java API has been left without standard classes for direct access to memory. This leaves Java at a distinct disadvantage to C and C++, which are able to do byte-level memory manipulation easily. Also, Java defines no unsigned integral primitive types, but this can be worked around without difficulty[1] since all the usual bitwise operators are available. In practice, any access to memory (usually for using memory-mapped I/O) must be done with native methods. Even ubiquitous hardware such as serial and parallel ports are unavailable from the core Java API. The javax.comm package contains a nice implementation for Win32 and Solaris, but has not been ported to other platforms.

13.2.3 Timing

Using native methods results in measurable application overhead, due to entry and exit from the virtual machine environment. Ironically, in order to even quantify this, a real-time clock must be read using native code. Although simple enough to implement, having to rely on a native interface for a service that needs to be as fast as possible is not desirable. Timing granularity varies depending on how the interpreter is run, but seems to be around 20-30 μsec when tested in an empty loop.[2] For simple native methods like reading a clock or accessing memory, this is slow but sufficient for many real-time applications. Ultimately, it is the responsibility of the user to implement accurate timing methods.

13.3 Advantages of Java for Control Software Design

Considering these challenges, one might ask why it is necessary (or even desirable) to write control software in Java. The simple answer is that control applications can benefit from all the design advantages Java has to offer without compromising real-time performance. Especially given the low-volume nature of many complex mechatronic control systems, working around some of Java's drawbacks may well be worth the benefits that it offers.

13.3.1 Modularity

Java offers the usual modularity that comes with object-oriented design. While this may not seem especially important for many embedded devices, it does become relevant any time the system specification changes. Any project which is completely specified at the outset, and that additionally will never need to be changed or upgraded, and further will never need to be worked on by more than one person, probably has no need for object-oriented design. Any other application, no matter how small the system, can benefit from the data encapsulation and code modularity offered by class-based languages. Control systems, in particular, are often made of several different modules which can be effectively described by classes (even if a task-based representation is not used). Keeping the code for these system modules

---

[1] The unicode char data type is unsigned, and can be effectively used as an unsigned integer value.

[2] This test was run on a PIII 500 processor under NT 4.0, using the Java 1.1.8 runtime from Sun.

separate and generalized allows for easier system upgradability and scalabilty, to say nothing of making it easier for the developers themselves.

13.3.2 Distributed Control

As microprocessors become both faster and less expensive, the idea of devices interacting in distributed networks becomes more practical. Using a network-enabled, dynamically loadable language such as Java to program these devices only makes sense. Bringing control code down to the device level, while at the same time making the devices smarter, flattens and simplifies the typical control hierarchy. The built-in portability of Java, combined with new technologies such as XML and SOAP for object communication,[3] suggests that control system components could possess a high degree of autonomy in the future. Some very real possibilities include system components that check for and perform their own firmware upgrades, device-level handling and alerting of exceptional system conditions, and truly plug-and-play components that can be introduced into a control network with minimal reconfiguration.

13.3.3 Platform Independence and Prototyping

The binary bytecodes read by the Java interpreter are inherently portable. This means that bytecodes produced on any platform should be readable by a compliant virtual machine on another. While it was this portability that initially made Java a popular language for web development, embedded software design stands to benefit even more in this regard. Considering that development seats and prototyping tools for embedded or real-time OS packages are typically expensive, the possibility of using only a commercial Java IDE for application testing and simulation is quite attractive. System control applications, in particular, can be substantially refined in simulation before ever running code on the target processor. Using commercial design tools like this can represent an enormous cost savings in development, as program flow bugs can be isolated in a relatively friendly environment.

13.3.4 Operator Interface Design

Probably the most appealing aspect of a pure Java user interface is the prospect of wrapping it in a downloadable form. Consider an embedded control system capable of delivering a Java applet to an operator using an arbitrary client. The resource bundle for this applet could reside on the embedded device itself, or perhaps on a central server. Depending on the system, a high degree of control could be available to the operator through a very sophisticated graphical interface. The client processor would perform the time-intensive graphical and event handling operations, freeing the embedded system to perform its tasks. Indeed, this hypothetical client system could be any processor that supports a virtual machine, so long as there was network connectivity and an established communication protocol. An excellent

---

[3] www.w3c.org/XML, www.w3c.org/TR/SOAP

human-machine interface (HMI) platform might well be a thin client with little more hardware than a processor, touchscreen, and an Ethernet connection.

13.4 Java and the Task/State Design Method

In the sense that Java is a class-based, object-oriented programming language, it is just as suitable as some others for implementing a scheduling package based on the task/state design method. It allows tasks and states to be realized as classes, from which object instances can be created and manipulated in the context of a control program. However, there are a few language features that are particularly helpful.

13.4.1 Inner Classes

As of Java version 1.1, one substantial difference in the language is the addition of inner classes. Instead of restricting public class declarations to one per source file, the new specification allowed multiple classes to be defined within each other. These inner classes could themselves be top-level, members of the enclosing class, or even completely local to a block of code. The relationship between enclosing classes and member classes is especially important, given that both classes have full access to each other's private fields and methods. This arrangement has the side effect of creating two separate hierarchies for classes: one of inheritance and one of containment[21].

This feature of Java lends itself particularly well to a software implementation of tasks and states. Although the two are naturally related in the physical design model, software classes representing tasks and states do not necessarily have to be related. Defining the base state class as a member of the base task class reinforces the relation between the two concepts by allowing the two classes to share methods and data. Additionally, user-defined task subclasses can enclose code for their own states, which is an excellent organizational tool. The use of inner classes is illustrated in the TranRunJ examples and chapter 17.

13.4.2 Networking

One of the principal features of the task/state design is the creation of intertask communication that does not involve direct method calls on other task objects. By requiring all communication between tasks to take place through API-defined method calls, the task or process class can implement the message passing however necessary. If multiple processes are being run on multiple machines, intertask communication necessitates both a network connection and an available network API. Java provides this in the java.net package, which offers clean and platform-independent networking classes. In TranRunJ, communication between tasks is implemented at the process level using a simple datagram protocol. Task messages and shared data are passed around using binary streams and datagram packets, but this is effectively hidden from the user.

The `java.net` package is convenient and easy to use, but a far more powerful technology for intertask communication could be Remote Method Invocation (RMI). This technology is based on a framework of local and remote objects, any of which can call methods of the others. These method calls are presented as normal object interaction, but this hides a layer of network connectivity and object reflection that actually calls the method on the remote object and gets the result. If this design were applied to intertask communication, it could result in a task-based system that was more peer-to-peer in nature, rather than having parent processes mediate the interaction between tasks. Unfortunately, Sun's RMI is not supported by many third-party real-time virtual machines, so for the present, any RMI-style solution would likely have to be custom written.

13.4.3 Documentation

Comprehensive documentation is critical to the management of any complex codebase. This is especially true for mechatronic system control applications, where the code corresponds directly to some physical system behavior. The task/state class hierarchy, combined with the class/package structure of Java, goes a long way toward giving context to modules of complex application code. However, just as in the past, documentation of the code is necessary in order to ensure maintainability. The Java language helps enable this by defining some comment syntax that allows properties of the source files to be easily extracted. JavaDoc comments relating to class function, methods and their parameters, inheritance and inner class hierarchy, and author or version information allow documentation to become an integral part of the application source code. Along with the Java SDK, Sun provides a utility that parses Java source files and generates documentation in HTML. Proper documentation of control code can offer tremendous time savings, given that any application will likely need upgrading at some point during the usable lifetime of the mechatronic system.

13.5 The Current State of Real-Time Java

As noted, in order to use Java in real time, the two necessary components are a deterministic virtual machine and a real-time operating system (RTOS). Additionally, there is a need for high-resolution timing and, almost always, the ability to write and read memory directly. At the time the TranRunJ project (discussed further in chapter 17) was undertaken, there was still not much offered as far as real-time Java solutions, but an acceptable combination was found, and the use of native methods allowed for timing and hardware interfacing. However, the native methods were often used to access necessary features of the RTOS, such as shared memory, and some generality of the scheduler had to be sacrificed.

This sort of VM/native methods/RTOS trio necessarily moves away from the touted platform independence of Java. In recognition of this, two independent groups have recently issued recommendations for a real-time Java platform. These include both an updated virtual machine specification and corresponding API extensions. Unfortunately, one group is sponsored by Sun and its Java Community Process[22], while the other has the backing of a number of Java licensees[23]. The two proposed APIs are not compatible, but hopefully some reconciliation will be possible soon[4]. Until then, real-time Java solutions will not be fully portable. The best option for portability seems to be careful isolation of code that relies on native methods or vendor-specific classes.

Possibly the most tightly integrated solution for the PC architecture at this point is the combination of the IBM J9 virtual machine and the QNX Neutrino RTOS.[5] J9 is a fully featured, mostly Java 1.2 compliant VM implementation, complete with networking support. In addition to running regular Java bytecodes, J9 also supports Sun's Java Native Interface (JNI) for native methods. It is also based on the specification advanced by the Java Community Process, which means that the real-time API extensions may very well still be supported in the future.

An interesting new real-time option is that of dedicated Java hardware. Recently, microprocessors have become available that use Java bytecodes as their native instruction set.[6] These exploit the existing Java API (generally the J2ME version), so that a minimal impact is made on code portability. One serious advantage of these processors is that the existing Java threading model can correspond directly to that of the hardware, without the complications of an intermediate OS layer. Rather than using Java as a normalizing wrapper for an operating system, a Java processor/runtime combination does away with the OS completely. Other recently introduced systems use the virtual machine as the RTOS, but port the software for different platforms.[7] Such systems have an advantage in that they use standard microprocessors rather than a specialized Java chip.

---

[4] As of November 2001, the Sun-supported effort has been finalized: http://www.rtj.org
[5] http://www.embedded.oti.com, http://www.qnx.com
[6] For example, the new AJ-100 from aJile: http://www.ajile.com
[7] For example, Jbed by Esmertec: http://www.esmertec.com

Chapter 14

PROGRAMMABLE LOGIC CONTROLLERS (PLCs)

14.1 Introduction

Programmable logic controllers (PLCs) have established a firm place as control elements for mechanical systems. Because their dominant programming model is ladder logic, based on relay logic, they are usually viewed as an entirely different class of device than computer-based controllers programmed in algorithmic languages such as C. The result is separate, somewhat hostile communities of people, those who use PLCs and those who use computers.

Through the use of a task/state design and implementation model for mechanical system control software (mechatronic software) these two seemingly disparate means of implementing mechatronic system control can be reconciled. Because the design model relates closely to its computational realization, control system designs can be transported between PLCs and computers relatively easily. This reconciliation is possible because the algorithmic software design is based on fully nonblocking code leading to a scanning computational structure. This structure is completely consistent with the rung-scanning algorithm used by PLCs.

Mechatronic systems are characterized by relatively sophisticated control implemented through computation. "Control" in this context covers the full gamut of control functionality: sequencing, operator interface, safety, diagnostics, and signal processing, in addition to feedback control functionality where needed. Two technologies to implement this control have existed side-by-side for a long time:

- Computers programmed with algorithmic languages

- PLCs programmed in ladder logic

They have been treated as independent technologies because their programming languages are so different that there tends to be little overlap among users. In particular, PLC ladder logic requires little of the computing infrastructure and training needed to build effective control (real-time) programs in common algorithmic languages such as C, C++, Pascal, Ada, and so on. It thus has great appeal to people who spend a lot of time working directly with the target hardware. On the other hand, algorithmic languages are more flexible than ladder logic and can be used in more sophisticated real-time environments (priorities, semaphores, etc.).

In both cases, however, the underlying computation is accomplished by a general purpose computer so there should be some hope of unifying these approaches and providing more effective choice for all mechatronic system designers.

Ladder logic has its basis in relay control systems. Solenoid-driven relays can represent Boolean variables, for example, value=1 for open, value=0 for closed. Combining these in serial/parallel combinations can implement Boolean logic equations (examples will follow). They were widely used to implement logic control systems in the era before widespread use of electronics or microcomputers.

Programmable logic controllers exploit the simplicity of relay logic systems by basing their programming language on a pictorial representation of relay logic, the ladder diagram. The relay logic concept is extended in ladder logic by generalizing the output action that can take place to include triggering of complex elements such as timers and even execution of short blocks of algorithmic code.

Control systems based on algorithmic languages owe their heritage to assembly language and to computer languages invented for the purpose of doing scientific computing and similar tasks (Fortran, Algol, Basic, etc.). None of these languages was invented for the purpose of interacting with physical systems on a timely basis. Therefore, using one of them or a derivative language in any control environment containing significant parallel operation of system components requires the addition of a support structure for some form of multitasking.

Another consequence of the isolation of PLC programming from the computer world is that system and control simulation is rarely done for projects using PLCs as controllers. There is no convenient interface that allows the PLC software to interact with a system simulation done in a standard PC or workstation simulation environment.

14.2 Goals

With this recitation of the problems associated with control system programming in general and PLC programming in particular, the following goals apply to this PLC programming methodology:

1. The design notation should allow for engineering review and design input from engineers familiar with the machine, not just those who are computer or PLC programming experts.

2. The process of producing functioning software from the design notation should be a translation process not a creative process.

3. It should be equally easy to produce PLC software, computer-based software, or a simulation from the design notation.

4. Relating sections of the target program to associated sections of the design document should be straightforward.

5. The PLC programming should utilize ladder logic.

6. No special software tools should be needed—just the standard target development environment.

The first four of these are general goals that lead to rational design and implementation. The last two are specific to the PLC world. The importance of retaining ladder logic as the low-level programming tool is because ladder logic remains the primary attraction of PLCs! The last goal is to make sure the design method is available regardless of the specific PLC being used and to preserve the nonproprietary nature of the design approach. The down side to the universality sought from goal #6 is that it remains up to the design engineer to keep the design documentation and the actual program in sync. If this happens, or if modifications are made to the ladder program that do not conform to the design model, the software complexity problems reappear.

14.3 PLC Programming

The PLC's natural domain is in problems for which all inputs and outputs are Boolean (two-valued) and for which the function relating the outputs to the inputs is algebraic (i.e., has no memory; is static or combinational). In this domain, PLC programs can be translated immediately from Boolean algebra to ladder notation without any loss of generality or fidelity. Internal variables can be used for convenience in this domain, but are never necessary.

14.3.1 When to Use a PLC

Unfortunately, very few real-world problems fit this model. Even if all inputs and outputs are Boolean, the control functions normally must have memory in order to meet the system control specifications. Such functions fall into the domain of sequential logic rather than Boolean logic. Implementation of sequential logic requires the use of internal variables to store historic information. These internal variables are often called state variables.

Sequential logic can be either synchronous or asynchronous. In synchronous logic input processing is separated from state variable update by a synchronizing clock. In asynchronous logic, these operations occur simultaneously. Sequence errors caused by seemingly insignificant changes in signal timing are difficult to avoid in asynchronous logic, leading to errors that can be probablistic and very difficult to even reproduce let alone track down. Design methods to eliminate or minimize these errors are quite tedious. Synchronous logic, on the other hand, is highly predictable and will be the recommended approach. While the original relay logic was asynchronous (relays respond immediately to inputs) ladder logic is executed by scanning and can be synchronous with proper design procedures.

Assuring that PLC programs are strictly synchronous has to move beyond the syntactical rules of ladder logic and embrace stylistic restrictions.

Common PLC usage extends beyond Boolean and sequential logic. When complex modules such as timers are used, or when arithmetic on real variables must be done, the Boolean or sequential logic is used to coordinate these activities. The further the problem strays from logic, the less useful ladder logic is as a fundamental organizing language.

14.3.2 Ladder Logic

Ladder logic, as noted, is a programming language based on hardwired logic using relays and coils[24][25][26]. It has been so long since that connection was made, and ladder logic has been modified so much from pure relay representation, that it can now be considered to be a pure programming language, an abstraction of its original physical model.

The basic programming unit, equivalent to a statement in algorithmic languages, is a rung (of a ladder). A rung has two parts: the left side is a Boolean expression, and the right side is an action based on the results of that expression. The action can take a number of forms. The rung in figure 14-1, for example, expresses the Boolean relation

```
x = (A AND (NOT B)) OR C
```

This is the purest form of ladder notation and follows exactly from relay logic. The symbol] [is for a normally open relay and the symbol]/[is for a normally closed relay. The name above the symbol is the associated variable. For the normally open case, the value appears as-is when translated into Boolean logic form; for the normally closed case, it appears complemented (primed, NOT) in equation form. The symbol ( ) indicates an output to be energized according to whether the rung is "conducting" (true), or "non-conducting" (false). Any number of rungs can be combined to make a ladder, which is equivalent of a function or a program in algorithmic languages. For example, the set of Boolean statements,

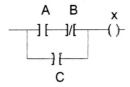

Figure 14-1. Basic rung of a ladder

Section 14.3. PLC Programming

```
x = A OR B
y = B AND (NOT C)
z = (NOT A) AND B
```

is represented by the ladder of Figure 14-2.

An extension of the basic ladder notation uses the concepts of latch and unlatch. The symbol (L) indicates that its associated variable will be set to 1 if the logic expression portion of the rung is 1 and will otherwise remain unchanged. Likewise, (U) indicates that the associated variable is set to 0 if the logic expression of the rung is 1 and unchanged otherwise. In this case the logic expression of the rung indicates whether or not an action should be taken. In the ladder of figure 14-3, for example, y is set to 1 if the expression (B AND NOT C) is 1; otherwise it retains its previous value. The variable z is set to 0 if ((NOT A) AND B) is 1; otherwise it is left unchanged. While the simple rung structure corresponds to a basic assignment statement in an algorithmic language, the latch/unlatch corresponds more closely to an if statement.

Function blocks use the rung's logic expression's value to indicate whether or not the block should be executed. In some cases, such as a timer block, the action is activated by a transition of the logic value from 0 to 1. Function blocks take the place of the result value on a rung as in figure 14-4.

Execution based on a transition can be made explicit (rather than implicit as it is with a timer) by using a one-shot-rising (OSR) block. With this block present, items following on the rung will only be executed if the preceding logic value has just done a 0-1 transition. In figure 14-5, for example, the latch operation is only done if the logic function (B AND (NOT C)) was 0 during the previous scan and is 1 during the current scan.

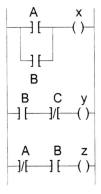

Figure 14-2. Boolean function in 3 rungs of a ladder

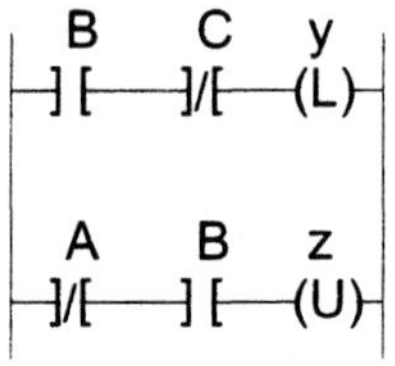

```
     B      C       y
    ─┤ ├───┤/├────(L)─

     A      B       z
    ─┤/├───┤ ├────(U)─
```

Figure 14-3. Ladder showing latch/unlatch

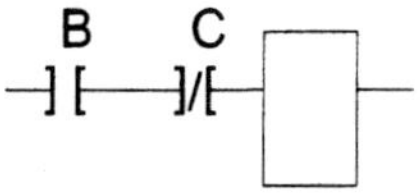

```
     B      C    ┌─────┐
    ─┤ ├───┤/├───┤     ├─
                 └─────┘
```

Figure 14-4. Function blocks

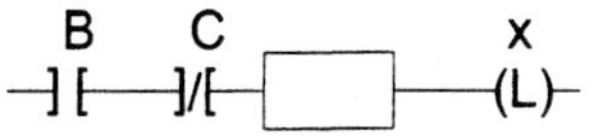

```
     B      C    ┌─────┐   x
    ─┤ ├───┤/├───┤     ├──(L)─
                 └─────┘
```

Figure 14-5. One-shot rising output

14.3.3 Grafcet/Sequential Flow Charts

Grafcet (old name, currently called sequential flow charts) is a method of sequential logic specification based on a modification of Petri nets[25]. It is commonly used to program PLCs. Once the logic has been specified using this method, ladder code is produced automatically to run the PLC. In the context of this paper, the unique feature of a PLC is its ladder logic programming. The ladder produced from a grafcet program is not accessible to the programmer so grafcet is thus a replacement for ladder programming and not relevant to the purposes of this chapter.

14.4 The Task/State Model

There are a number of similarities in the task/state model to PLCs. As presented earlier as a model for building real-time control software, the most striking similarity is that the nonblocking code restriction makes software constructed this way have very similar execution characteristics to ladders, which are scanned rung-by-rung. The corresponding execution structure for task/state-based algorithmic programs is that the code associated with the current state of a task is scanned repeatedly. We will exploit this similarity to show how control programs specified in task/state form can be translated to PLC-ladder notation as easily as to algorithmic code. This assumes that the problem is suitable for PLC implementation, however. That is, the bulk of the control decisions are made on the basis of binary sequential logic.

14.5 State Transition Logic for a PLC

For problems that are purely sequential logic, there are formal design methods that can be used to create PLC programs since ladders are formally equivalent to Boolean logic. These methods will not be pursued here because many (if not most) PLC problems contain some nonlogic elements and because using these methods would require that PLC users learn several methods of formal logic such as Karnaugh maps, next-state tables, and so on. For large problems, auxiliary software would also be required to implement the designs[27].

As noted, the design problem is simplified if it is restricted to synchronous logic. Because there is a single processor underlying the logic that scans the PLC rungs in sequence synchronous logic can be implemented. For synchronous operation, the design method must assure that any group of rungs constituting a logic unit have no order dependence—i.e., reordering the rungs does not change the results.

14.5.1 State Variables

State within a task is usually represented by an integer value. It serves to notate the current activity for the task. In algorithmic programming the state variable for each task is an integer. In ladder programming each variable is a Boolean (1-bit) so a set of Boolean variables is used to create a binary integer value. We will use the variable set ¡S1-n,...,S1-2,S1-1,S1-0¿ for state, with S1-0 for the lowest order bit for task 1, S1-1 for the next-to-lowest for task 1, and so on. Each task has a similar set.

14.5.2 Ladder Organization

There are a number of possible ways to organize the logic associated with a state. The recommended method collects all of the rungs for each state together so the connection between the ladder and the documentation remains as strong as possible. A single introductory rung tests the state—parallel rungs branch from that for the entry, action, test and exit activities.

The rung of figure 14-6 shows a typical example, in this case for the state 101 (state #5) of task #2. This example uses three state variables, allowing for a total of eight.

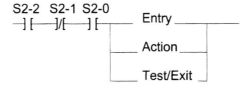

Figure 14-6. State representation in ladder

Specification of output variables correctly is critical to avoid ambiguity of output values. Outputs specified by simple energize operations (-()-) can be ambiguous if there are entries for the same output in several states so latch operations are used most of the time.

14.5.3 Transitions

As a task proceeds from one activity to another its state changes. These transitions depend on what state in the task is currently active and values of other variables—typically from sensors or from outputs associated with other tasks. Each state can have any number of potential transitions. These transitions appear in the test section shown earlier and provide the connection between the topology of the state diagram and the PLC code. The transition tests use latch/unlatch operations so the state variables are only changed if the test is TRUE.

Two transitions from state 101 (#5) are shown in figure 14-7. The first will go to state 110 (#6) if the expression (I1 AND I2) is TRUE. I1 and I2 are input our other variables. The second will go to state 100 (#4) if (I1 AND (NOT I2)) is TRUE. If neither of these conditions is TRUE, there will not be a transition. The variable **tran** is latched indicating that a transition is about to take place. It is used to perform state clean-up activities.

The use of variables with names that are different from the state variable names already introduced is important here. The n in front of the name stands for "new" (or "next"). The new state variable values will not be copied until all of the processing for this task is completed. This is the primary guarantee of synchronous operation. Any subsequent use of the state variables will use the existing state not the next state, thus preventing incorrect outputs or multiple transitions in a single scan.

If more than one transition condition is positive at the same time, the transition will go to the state associated with the last one (because of the scanning order of the rungs). Note, however, that if exit operations are involved this can lead to multiple

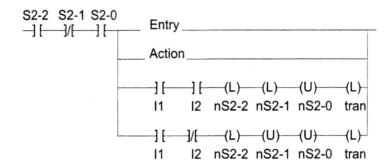

Figure 14-7. State transition

(and possibly incorrect) execution of exit outputs. For that reason, it is safest to make sure that transition conditions are mutually exclusive.

14.5.4 Outputs

Output activity can be associated with entering a state, entry, being in a state, action, or leaving a state, exit. The exit activity can be associated either with a specific transition or can be common for all transitions.

There are at least two ways to handle logic outputs whose values are associated with state. The method recommended here is to latch the value at the entry to the state and unlatch it on leaving (or vice-versa for an output with value 0).

14.5.5 Entry Activity

Entry activities must be triggered by a transition in state. This can be accomplished implicitly for elements such as timers that are sensitive to transitions or explicitly with an OSR element. If the OSR element is combined with an energize-type of output, an output pulse one scan wide will be produced each time the state is entered. A latch (or unlatch) operation on an entry rung will determine the value of a variable for the duration of the state. Figure 14-8 shows a ladder with one entry item of each type.

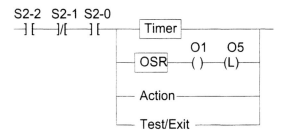

Figure 14-8. Types of outputs

14.5.6 Action Outputs

The action section is used for activities that must be done repeatedly while the task is in a state. These could include feedback control (e.g., testing the value of a comparator and setting an actuation output accordingly), servicing a sensor, and so on. It is common for this section to be empty.

14.5.7 Exit (Transition-Based) Outputs

Like the entry activities, these depend on a transition. In this case, however, it is a transition of the test condition. Figure 14-9 shows two such exit outputs. The output O2 will pulse whenever the transition based on I1,I2 = 11 is taken; O3 will pulse for the I1,I2 = 10 transition.

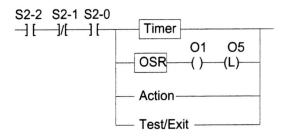

Figure 14-9. Transition-based outputs

14.5.8 Common Exit Activities

The primary usage of this section is to reverse the actions done in the entry section—for example, turning off outputs that were turned on. The execution of this section depends on the variable `tran` that is set if a transition is about to occur (it must also be cleared in this section). Figure 14-10 shows a common exit rung which turns off the output O5 that was latched in the entry section and then unlatches `tran`.

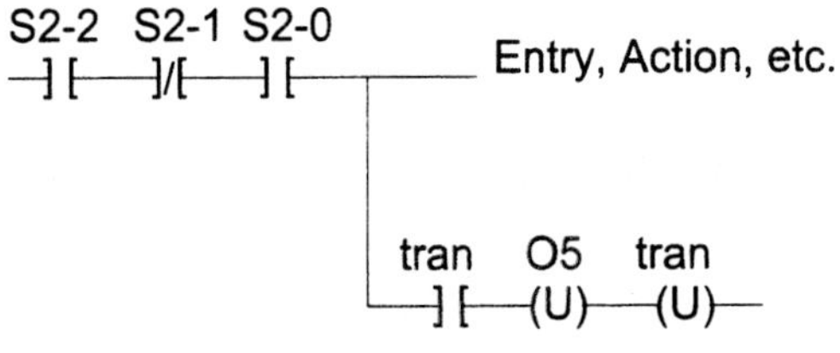

Figure 14-10. Reversing entry actions

14.6 PLC Multitasking

Equal priority (cooperative) multitasking can be accomplished quite naturally with a PLC by grouping all of the rungs associated with each task together. The PLC scan cycle will visit each task in turn. Some PLCs also offer various interrupt abilities from which preemptive multitasking can be constructed. This is only necessary when severe latency constraints are encountered.

14.7 Modular Design

Design modularity is achieved through the original task/state organization. The ladder program retains this modularity by grouping all rungs associated with a state in the same place. Thus functional bugs can be traced quickly to the responsible code. It is most important to note again, that changes must be made first to the design documentation and only then to the ladder program. If this rule is not followed, the design documentation quickly becomes obsolete and thus worse than useless—wrong!

14.8 Example: Model Railroad Control

A model railroad (HO gauge) installed in our teaching lab serves as the basis for this sample problem. The model train is set up to emulate an industrial system that operates primarily on logic sensors and actuators so is suitable for PLC control.

Magnetic proximity detectors are used to sense train presence via magnets mounted on the trains. The proximity sensors serve to isolate track blocks (figure 14-11). Actuation is through three power-isolated track sections and solenoid switch actuators.

The sample problem shown in figure 14-12 consists of topologically concentric, independently powered tracks, connected with switches. Each track has one train running on it. The control problem is to switch the train on the inner track to the outer and vice-versa, using the two switches so the maneuver can be done safely.

The control for this consists of four tasks: outside track control, inside track control, switch control and the "main" task. Figure 14-13 shows the transition logic for the "main" task, which controls the general switching procedure. When a reversal is requested each train is brought to a station just before its respective switch and held there. When both trains are in place, the track reversal takes place and then normal operation resumes. The ladder code for this problem can be found at: http://www.me.berkeley.edu/faculty/auslander/res.html: scroll to "Reconciling PLCs...."

The tasks for the track control are similar; the task for the inside track is shown in figure 14-14.

The final task controls the operation of the switches. Both the track control and the switch control tasks respond to commands from the main task, assuming

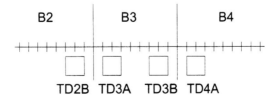

Figure 14-11. Magnetic train-detection sensors

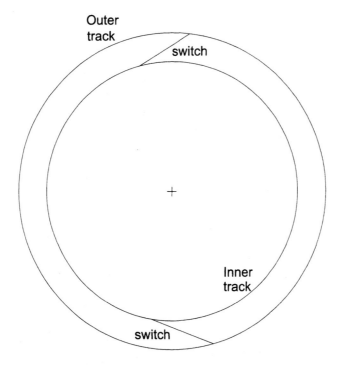

Figure 14-12. Concentric model railroad tracks

consistency with the local operating conditions. The switch control task is shown in figure 14-15.

14.9 Simulation – Portability

To demonstrate code portability for PLC designs done in this format, the sample problem is also presented in Matlab form. While Matlab has a general algorithmic language and thus is a stand-in for a number of potential control languages, this exercise also demonstrates the ability to simulate the performance of a target system from the same design documentation used to program the PLC—something that is very difficult to do with a PLC. Figure 14-16 shows the simulation results. The graphs can be interpreted as follows: the top graph is the position of one of the trains (angle vs. time), the bottom the other. Angles between 0 deg. and 360 deg. represent positions on the inner track; between 360 deg. and 720 deg. represent positions on the outer track. The jumps at time equal to about 170 are the trains switching tracks. Other jumps are just rollovers at 0/360 for the inner track and 360/720 for the outer track. The full Matlab programs are available at the web site referenced above.

Section 14.9. Simulation – Portability 233

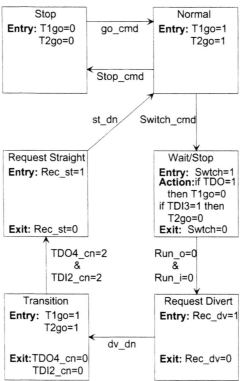

Figure 14-13. Main task

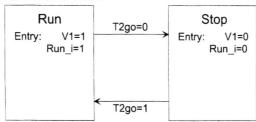

Figure 14-14. Inside track task

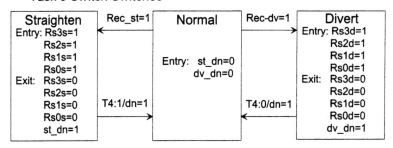

Figure 14-15. Switch task

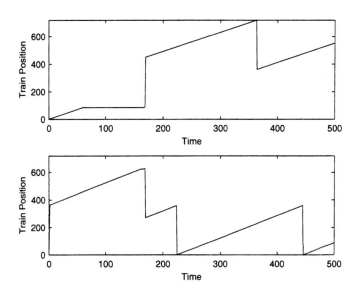

Figure 14-16. Train simulation results

Chapter 15

ILLUSTRATIVE EXAMPLE: ASSEMBLY SYSTEM

The example of designing and implementing the control system for assembling a simple object as part of a manufacturing process is presented in a step-by-step fashion. It is intended for use either as a fully worked out example or as a sequence of self-study exercises for which the example programs serve as solutions. This chapter used the group-priority software for building the example. chapter 16 shows a similar example using the TranRun4 software, and chapter 17 describes a Java version with TranRunJ.

15.1 The Assembly System

The object to be assembled consists of a square base to which a cylindrical object is to be glued (figure 15-1).

The assembly process consists of placing the square base into the assembly area, putting a thermosetting glue on the base, then placing the cylinder on top of the base. While there is some adhesion between the base and the cylinder at this point, it remains weak until the glue has been thermally processed.

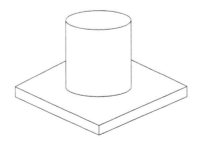

Figure 15-1. Object to be assembled

The base-cylinder is then moved to an oven where it is held at the thermal setting temperature for a specified time period. Then it is moved to an unload area where it is picked up for use in the next part of the manufacturing process.

Because the heat treatment is the slowest part of the process, a system with a single assembly area serving two ovens is used (figure 15-2).

The assembly area is served by two robot-like devices. One places both the base and the cylinder onto the assembly platform and the other applies the glue to the base.

The conveyor systems are driven by DC motors so speed and acceleration can be controlled. This allows for delicate handling of the assembly object in the first stage while it is moving to the oven and faster handling in the second while it is being moved to the unload area. The conveyors each have a single clamp which is used to attach the object to the belt. When no object is clamped to the belt, it can be moved independently.

The ovens are electrically heated and have temperature sensors. The goal is to maintain constant temperature in the oven all of the time.

The unload areas are each serviced by robots that move the finished assembly to the next processing area.

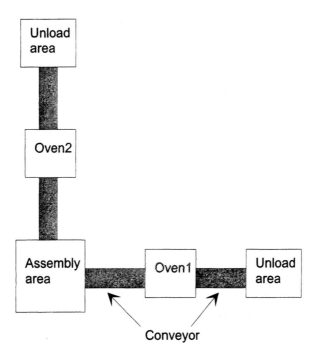

Figure 15-2. System to assemble glued objects

15.2 System Simulation

In order to be generally useful, the control system software design will be targeted to a simulation of the assembly system. A fairly detailed model will be used for the conveyor motion (including a servo control for the motor input-position output loop). The robots, however, are considered as devices which are controlled by separate control systems, so the simulation only includes a basic command set and delay times for various actions (more detailed models could be added).

The Euler integration algorithm is used for components described by differential equations (motion and temperature). This keeps the software simple at the expense of requiring small step sizes for the integration.

The simulations are represented by tasks in the control system, so are programmed in exactly the same way as the control tasks.

15.3 Development Sequence

The control system development is represented as a sequence of exercises. Each exercise results in a working program, generally building on and modifying the program that resulted from the previous exercise. A solution (that is, a working program) is given for each exercise. There is considerable duplication from one solution to the next, but to keep the results for each exercise independent, all of the files needed for each solution are kept individual directories, one for each exercise.

These solutions are certainly not the only possible solutions. What use is made of them depends on how this chapter is being approached. When used as a set of assignments, with no prior reference to the solutions, it is likely that the resulting software will look quite different from the solutions that are presented. In this case, the best way to use the given solutions is to examine them after each exercise has been completed to see where the given solution has features that could be used. An improved solution can then be constructed using the best of both.

The exercises can also be approached by studying and using the solution programs first. Then various modifications can be made to test and broaden understanding of the issues of each of the exercises.

The order of exercises does represent part of the solution in the sense that it establishes the task structure. An alternate exercise for advanced readers would be to take the "clean room" approach and start only with the problem statement. The solution arrived at can then be analyzed relative to the one given here.

Top-down versus bottom-up design is a much discussed subject. The approach here is a mixture. It is top-down in the sense that the entire design is done in simulation and can be used as part of the design process, before any part of the machine is actually built. On the other hand, the sequence of exercises is bottom-up. It starts with the simulation of the core hardware of the machine and then works up to the most abstract levels of the control.

The sequence of programs are all in the directory Glue, with each step in the sequence having its own directory, Glue00, Glue01, and so forth. This solution is done using the group-priority software base for C++ (archive grp_cpp.zip).

15.4 Belt Motion Simulation (Glue00)

The first program in this sequence is used to simulate the motion of the belts that move the assembly objects through the oven and to the unload area.

15.4.1 Modeling Belt Dynamics

The model used for belt motion is that of a command signal to an amplifier that produces current to drive a motor. The command is thus proportional to torque delivered by the motor to move the belt. The belt is modeled as an inertial load. The units of belt motion are in meters. Although suggested units are given in the comments of the program, no attempt at parameterizing the internal model has been made, so the units for current, torque, and so on are arbitrary. The equations for belt motion as used in the simulation are thus:

$$\tau = K_i i, \quad \frac{dv}{dt} = \frac{\tau}{J}, \quad \frac{dx}{dt} = v$$

where τ is torque, J is the system inertia, and x and v are position and velocity of the belt.

The Euler integration algorithm approximates the right-hand side of the differential equation as constant over the integration period and equal to the value at the beginning of the period. Given a set of coupled differential equations of the form

$$\frac{dx_1}{dt} = f_1(x_1, x_2, ..., t), \quad \frac{dx_2}{dt} = f_2(x_1, x_2, ..., t), \quad ...$$

and given initial conditions for all of the dependent variables, each step of the Euler integration is:

$$x_1(t + \Delta t) = f_1(x_1(t), x_2(t), ..., t)\Delta t + x_1(t)$$
$$x_2(t + \Delta t) = f_2(x_1(t), x_2(t), ..., t)\Delta t + x_2(t)$$
$$...$$

A small variation of this comes from noticing that in the sequence of calculations the value of $x1(t + \delta t)$ is known before the x_2 is carried out. Although it may look more complicated, the following form is simpler to compute and can be more stable than the original form:

$$x_1(t + \Delta t) = f_1(x_1(t), x_2(t), ..., t)\Delta t + x_1(t)$$
$$x_2(t + \Delta t) = f_2(x_1(t + \Delta t), x_2(t), ..., t)\Delta t + x_2(t)$$
$$...$$

For the motor-driven belt system, this can be coded as

```
torque = Kt * cur;
v += (torque / J) * delt;
x += v * delt;
```

These three lines form the heart of the program that will be created for this step of the system design. The purpose of this program is to simulate the operation of the belt axis motion. No control or other system components are present. As usual, however, the three code lines that express this functionality are swamped by the rest of the code that is used to make them work.

The process of creating tasks will be outlined here in some detail. It is the same for all of the subsequent steps, so will not be repeated except to point out additions or changes to suit specific program needs.

15.4.2 Definition of Task Classes

The process begins with specification of the tasks that will be needed to meet the functional needs.[1] In this case, there are three tasks needed:

- Simulation of belt axis #1

- Simulation of belt axis #2

- Data logging

The first two are fairly obvious. The third task serves to separate the process of archiving of data for later analysis from the actual simulation process. This allows independent optimization of computing needs for these functions and also provides for the data logging function to gather information from a number of tasks before sending it to the archive medium (usually a disk file).

The next decision is to relate these tasks to the computing language environment that will be used. This program will be done in C++. The two simulation tasks are very similar, so it is necessary to decide how much C++ class definition can be shared. In this case, the two axes are assumed to be physically the same, so a single class will be used to instantiate both tasks.

The task implementation process starts with the tasks.hpp file. This file contains all of the class definitions for the system and is included in all of the files constituting the program. Every job has a tasks.hpp file.

The top part of tasks.hpp doesn't change much from job to job. It pulls in the include files for general use (iostream.h, etc.) and those that are unique to control jobs. Of these, time_mod.hpp is of particular importance because it is used to set the timing mode used by the program. This is the only one of the include files that is normally modified by the programmer.

Just below that is some material relating to process definitions. It is set up to allow for later manipulation of process usage, but to offer minimum interference with initial programming using a single process.

The next section is where the classes are defined from which tasks will be instantiated. Tasks used for simulation are separated by an #ifdef preprocessor directive so that real tasks can be subsituted easily.

---

[1] tasks.hpp

The first class defined is for data logging. Its class definition is typical of most such definitions, although simpler since this task is not very complicated. The class definition is:

```
class CDataLogger : public BaseTask
  {
  public:
     CDataLogger(char *name,int ProcNo=0);  // Constructor
     ~CDataLogger(void); // Destructor
     int Run(void);
     void WriteFile(char *filename);

  private:
     DataOutput *DataOut1;
     double x1,v1,x2,v2;   // Data to be logged
  };
```

All task-related classes inherit their basic properties from BaseTask (its details can be found in the files basetask.hpp and basetask.cpp). The public interface must always include a function named Run(). This function is called to execute the task-based code. Any other public interface elements can be included as needed. The private section consists of data and function as needed by the specific task.

The class definition for the belt axis is similar:

```
class CBeltAxis : public BaseTask
    {
    public:
        CBeltAxis(char *aName,double JJ,double Ktt, double ccur,
              int ProcNo);// Constuctor for the class. Put in parameter
                     // values here so the various instances can be
                     // distinguished
        ~CBeltAxis();     // Destructor for the class.
        int Run(void);
    private:
        double x,v;  // Position and velocity. Position is 0 when the clamp
              // is at the part pick-up point.  m, m/s
        double cur;  // Current to the motor, amps
        double J;  // System inertia, as viewed from the motor, N m s^2
        double Kt; // Conversion from motor current to torque, N m /amp
        double previous_time;  // Used to compute the time increment for
              // the Euler integration
        };
// Define global variable indexes
#define BELTAXIS_x 0
#define BELTAXIS_v 1
```

Several arguments have been added to the constructor. This is used so that when several tasks are instantiated from the same class, the parameters can be customized for each task. The private variables defined are used by the simulation. Following the class definition the indexes for message boxes and global variables are defined. In this case, there are no message boxes. The global variables are defined so that the data logger can get access to the data it needs. Note that by putting these definitions here they become accessible to all other files in the program.

Some function prototype definitions come next. The ones in this file are common to all applications; additional prototypes can be added if some functions are defined for general use which are not class member functions. There are none of that sort in this case.

The next section is used to declare a set of global pointers that will be used to access tasks. To properly use the **extern** property of C++, the declaration must be different in the file that "owns" the global variables. This is done by using preprocessor macro based on whether the statements are being compiled within the main file (identified as MAIN_CPP). The effect of this is to make these pointer variables available anywhere in the program.

```
#ifdef MAIN_CPP
    #define EXTERNAL   // So externs get declared properly

#else
    #define EXTERNAL extern
#endif

// Tasks
EXTERNAL CDataLogger *DataLogger;
#ifdef SIMULATION
    EXTERNAL CBeltAxis *BeltAxis1,*BeltAxis2;
#endif
```

These statements define the task pointers, but the tasks themselves have not been instantiated so the pointers are not pointing to anything yet. That situation will be remedied in the main file.

The final section defines lists that are used to keep track of the tasks. These definitions are the same in all jobs.

15.4.3 Instantiating Tasks: the Main File

The main file is where all of the operations necessary to start the program and then to do an orderly shutdown are done. Most of that work is done in the function that is called when the program first starts. In the C++ console mode, this is normally a function called **main()**, but other names may be used for different environments. The file name for this program is GlueMain.cpp.

The top part of the file and the beginning of **main()** are generic items that are present in every job. Task creation (instantiation) is the first program section that is job specific. The new operator is used to instantiate tasks and initial parameter values are passed to the task via the constructor:

```
// Create all of the task objects used in the project
DataLogger = new CDataLogger("Data Logger",PROCESS_A);

// Define task objects for running the simulation.
// These tasks are scheduled just as any other tasks, and are
```

```
// only added to the task list if a simulation is desired.
// The actual simulation code in contained in the Run()
// member functions.
#ifdef SIMULATION
BeltAxis1 = new CBeltAxis("BeltAxis1",25.0,4.0,3.0,PROCESS_A);
BeltAxis2 = new CBeltAxis("BeltAxis2",25.0,4.0,2.0,PROCESS_A);
#endif //SIMULATION
```

The result of the new operator is a pointer to the object (task) that has been instantiated. That value is assigned to the pointer variables that are defined in tasks.hpp. Note that the same class, CBeltAxis, is used to instantiate the tasks for both belts.

Once tasks have been instantiated they must be added to appropriate lists so that they will get executed. The programming model used here only requires that a task be added to such a list; the programmer does not have to do anything further to assure that the task is executed.

For this program all of the tasks are added to the same list, the low priority list. The if statement concerning which process a task is in is present here so that multiprocess solutions can be constructed easily at a later date. For this program, all tasks are in PROCESS_A.

```
if(ThisProcess == PROCESS_A)LowPriority->Append(DataLogger);

#ifdef SIMULATION
if(ThisProcess == PROCESS_A)LowPriority->Append(BeltAxis1);
if(ThisProcess == PROCESS_A)LowPriority->Append(BeltAxis2);
#endif //SIMULATION
```

The remainder of the main file contains material that is generic to all jobs (except for the request to write the data logging file).

15.4.4 The Simulation Task

The code associated with tasks is normally collected in one or more files, separate from the main file. For this program all of the simulation tasks are in the file Sim.cpp. The belt axis simulations are tasks that do not need any internal state structure. That is, each time the task is scanned the same code is executed. They are otherwise similar to any other tasks, so the file description here will be a model for any other tasks that do not use internal structure.

The primary purpose of the constructor is to set initial values for parameters and variables, as shown next. It also sets up the array of global variables with the call to CreateGlobal(). The call to ActivateProfiling() is commented out here because no actual real-time calculation is being done at this time—only simulations. When real time is used (including real time simulation) it gives information on how much computing time each task uses.

Section 15.4. Belt Motion Simulation (Glue00)

```
CBeltAxis::CBeltAxis(char *aName,double JJ,double Ktt, double ccur,int ProcNo)
        :BaseTask(aName,ProcNo)
{
DeltaTaskTime = 0.1; // Nominal task period, sec.
NextTaskTime = 0.0;
State = 0;         // set to initialize
previous_time = 0.0;
CreateGlobal(5); // Create a global data base with 5 slots
// ActivateProfiling(5.e-6,80.e-6,15,LOG_PROFILE); // Not needed yet
// Set parameters
J = JJ;
Kt = Ktt;
cur = ccur;
// Set initial conditions
x = 0;  // Position and velocity
v = 0;
}
```

No destructor is needed for the belt axis simulation task because it does not create any objects that need to be deleted on exit.

A task's Run() function gets executed repeatedly (scanned) once regular operation is started. The first part of Run() contains code that is used to manage the internal structure (states). Since this task does not use any internal structure, only a vestigial part of that code remains.

The actual simulation code is run at a time interval of DeltaTaskTime. It advances the simulation one time step and copies information to the global variables,

```
    ...
    CBeltAxis::Run(...)[ ]
    Time = GetTimeNow();
    // Check for time to do the next simulation computations
    if(Time >= NextTaskTime)
        {
        NextTaskTime += DeltaTaskTime;

        double delt;

        Time = GetTimeNow();
        delt = Time - previous_time; // Compute actual time
                // increment for Euler integration
        previous_time = Time;

        // Start belt axis simulation
        double torque = Kt * cur;
        v += (torque / J) * delt;
        x += v * delt;

        // Copy data to local arrays
        PrivateGlobalData[BELTAXIS_x] = x;
        PrivateGlobalData[BELTAXIS_v] = v;
        CopyGlobal(); // Make data available to other tasks
    ...
```

15.4.5 The Data Logging Task

The data logging task file, `datalog.cpp`, has the same general structure as the simulation task so only the aspects of it that are unique to the data logging function will be highlighted here. Data logging is handled by a class that is part of `basetask.hpp/cpp`. It is designed to keep data in memory until a specific request to write to a file is received. This allows for minimum interference with real-time operation since file operations can be quite time consuming, sometimes in unpredictable ways. By handling all of the details of data logging as a utility, the programmer only has to provide a minimum amount of code. The main cost for this is flexibility: the current configuration for data logging only allows for arrays of doubles to be logged. Any number of different logging functions can be handled simultaneously, however (up to the limits of memory).

The constructor is a convenient place to create the data logging objects — as a general rule it is a good idea to restrict operations that will require memory allocation to the portion of the program that runs before control operation starts. Memory allocation can consume significant amounts of processor time. Data logging objects belong to the class `DataOutput`.

```
CDataLogger::CDataLogger(char *name,int ProcNo)   // Constructor
    :BaseTask(name,ProcNo)
  {
  DataOut1 = new DataOutput(4,0.0,20.0,0.1);

  // Initial values for output variables
  x1 = x2 = v1 = v2 = 0.0;
  State = 0;  // For initialization on the first run
  }
```

The parameters for the `DataOut` constructor are:

- Number of values to be logged

- Time to start logging

- Time to end logging

- Time interval for logging

The destructor for the class is used to delete any objects created,

```
CDataLogger::~CDataLogger()// Destructor
  {
  delete DataOut1;  // Recover memory used
  }
```

Usage of the data logging object is in the `Run()` function,

```
if(DataOut1->IsTimeForOutput())
  {
  // Get new values
  CriticalSectionBegin();  // To make sure values are consistent even
```

Section 15.4. Belt Motion Simulation (Glue00)

```
                // if preemptive scheduling is in use
x1 = GetGlobalData(BeltAxis1,BELTAXIS_x);
v1 = GetGlobalData(BeltAxis1,BELTAXIS_v);
x2 = GetGlobalData(BeltAxis2,BELTAXIS_x);
v2 = GetGlobalData(BeltAxis2,BELTAXIS_v);
CriticalSectionEnd();
DataOut1->AddData(x1,v1,x2,v2, END_OF_ARGS);
```

The `if` statement checks to see if it is time to log data. The actual test is handled by the data logging class itself. If so, data is gathered by getting global information from the tasks. In this case, the two belt axis tasks have the needed data. Note the use of the pointers to the tasks that were originally declared in `tasks.hpp` and then instantiated in `GlueMain.cpp`. The `CriticalSectionBegin/End` function calls that surround the data gathering assure that all four data values will be from the same sample set so data consistency can be maintained.

Adding data to the log is accomplished with a call to the class member function `AddData()`. The `END_OF_ARGS` value is needed for proper operation (`AddData()` uses a variable length argument list).

The data log file is written when the real-time (or simulated) operation is over. It is done with the following line in the main file (`GlueMain.cpp`):

```
if(ThisProcess == PROCESS_A)DataLogger->WriteFile("gluedata.txt");
```

The `WriteFile()` function is defined in `datalog.cpp`, which accesses the output method in the `DataOut` class.

```
void CDataLogger::WriteFile(char *filename)
  {
  DataOut1->WriteOutputFile(filename);
  }
```

15.4.6 Timing Mode

The kind of timekeeping a program will use is determined by the settings that are made in the file `time_mod.hpp`. All of the initial samples in this series are purely simulation so use the internal timing mode. In internal timing, a variable representing (simulated) time is incremented as the calculation proceeds. This is done by selecting the `TIMEINTERNAL` option and by commenting out all other options in `time_mod.hpp`,

```
 #define TIMEINTERNAL // Use internal time calculation only
// (i.e., calibrated time). TICKRATE sets the internal
// time increment.
```

The time increment used is determined by the internal value of a `TickRate` variable. It can be set by a define in the `time_mod.hpp` file,

```
#define TICKRATE 0.01
```

or by a function call in the main file (`GlueMain.cpp`).

15.4.7 Compiling

A variety of compilers have been used with these files from time-to-time—Microsoft Visual C++, Borland C++ (v5), Borland CBuilder, GNU C++, and so forth. Project and workspace files are included in the archives for Borland C++ and Microsoft Visual C++. However, it is difficult to make these project files independent of the directory structure on the computer being used, so they should be approached with caution.

Many of these examples do not use operator interfaces or network access so they do not need all of the files in the Common directory. The files actually needed are:

```
AUDIT.CPP
AUDIT.HPP
BASETASK.CPP
BASETASK.HPP
CLOCK.CPP
CLOCK.HPP
COMPCONV.HPP
```

Unless noted otherwise, all of the samples are compiled as 32-bit executables targeted as console applications (that is, they use the text-based prompt window for execution). No other special compiler options are used.

15.4.8 Results

The program executes in a console window (also called an MS-DOS Prompt window, although it is not MS-DOS that is executing). It writes progress and debugging messages to that window using the cout facility of C++ (user data can be mixed into this stream also). The first part of the output shows the tasks that have been registered with the various lists for execution,

```
Added "Data Logger" as task: 1 to task list "Low Priority"
Added "BeltAxis1" as task: 2 to task list "Low Priority"
Added "BeltAxis2" as task: 3 to task list "Low Priority"
Maximize window if necessary. ENTER to continue...
```

When the program finishes it indicates why it stopped (a normal exit in this case) and then gives some summary statistics,

```
***Normal Program Termination***

The time at termination is 2
Task            #Scans

Data Logger         60
BeltAxis1           60
BeltAxis2           60

Total scans, all tasks: 180


Total running time: 2
Hit any key to exit.
```

The data generated from the data logger is in the file `gluedata.txt`. It is shown in Figure 15-3 in graphical form (plotted in Matlab using the script file `plotbelt.m`).

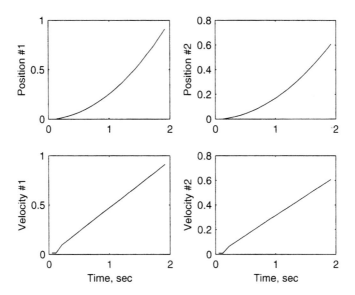

Figure 15-3. Belt axis simulations results

15.5 Oven Temperature Simulation (Glue01)

Simulation of the thermal process in the ovens follows exactly the same pattern as simulation of the belt motion. One class is defined and two tasks are instantiated from it, one for each oven. The code that does the actual simulation is:

```
double HeatIn = Cur * Cur * R;
double HeatOut = K * (Th - ThAmb);
Th += ((HeatIn - HeatOut) / HCap)* delt;

// Copy data to local arrays
PrivateGlobalData[THOVEN_Th] = Th;
CopyGlobal(); // Make data available to other tasks
```

This code is located in the `Run()` function for the class `CThOven` in the file `sim.cpp` along with the corresponding code for the belt axis simulation.

15.6 PID Control of Belt Position and Oven Temperature (Glue02)

Both the belt axes and the ovens use feedback control – position control for the belts, temperature control for the ovens. A PID controller will work for both of these. A PID control class is defined here that is similar to the one used in other examples, but uses different means for some of its data exchange.

15.6.1 Keeping Classes Generic

The reason different data exchange means had to be devised is that the target systems being controlled each instantiate two tasks from the same class. The method used to communicate the process variable value and the actuation output between the object being controlled and the control task was via virtual functions. Each PID control task that was derived (by inheritance) from the base PID class inserted its own version of these virtual functions. In this case, however, there are two identical sets of PID control tasks to create, one for each of the belt axes and one for each of the ovens. Each of these two controller tasks has its own unique data set, but it shares the same code and thus shares the same virtual functions. However, the virtual functions were where the code that was unique was located. That would not work for this problem. While ways to work around this could be devised, it seemed cleaner to look for a method that would make the PID control class more general.

This was done by using a mixture of message boxes and global variables. The control code needs to know how to get the process variable. If the process variable is represented by a global variable, the question is how does the control code know which global variable is in which task. This is solved by passing that information to the control task via its constructor. The arguments for the constructor include the task and the global variable index for the process variable. Likewise, the control task needs to know where to put the actuation value. Global variables cannot be used for push operations, but message boxes can. Again, the constructor is used to establish the task and message box number that the controller can use to send out the actuation value.

With this scheme, the virtual functions are no longer needed so it is possible to have multiple control tasks by just instantiating them all from the same class. All four control tasks are instantiated directly from the `PIDControl` class.

In order to make the PID control material maximally usable, it is contained in separate header and code files, `pid_gen1.hpp` and `pid_gen1.cpp`. The header file is included in `tasks.hpp` to make it accessible.

15.6.2 The PIDControl Class

The class presented here shows how the process and actuation variable information appears in the class and in the argument list of its constructor:

```
class PIDControl : public BaseTask
{
    protected: // This is just the generic part of a control task
    // so all variables are made accessible to the derived class
    double integ;
    double set,val;      // Setpoint and output (position) value
    double prev_set,prev_val;
    double kp,ki,kd,min,max; // Controller gains, limits
```

Section 15.6. PID Control of Belt Position and Oven Temperature (Glue02)

```
    double dt; // Controller sample interval
    double mc;      // Controller output
        int first; // Determines if PID is just starting
        BaseTask *ActuationTask;  // Task that will get the
            // Actuation value
        int ActuationBox; // Box number to put actuation value into
        BaseTask *ProcessTask; // Task from which to get proc. value
        int ProcValIndex; // Global data index for process value
        int ControllerOn;
  public:
    PIDControl(char *name,double dtt,
               BaseTask *ActTask,int ActBox,
               BaseTask *ProcTask, int PrIndex,
               double aKp,double aKi,double aKd,double aMin,double aMax,
               double aSetpoint,int aControllerOn,
               int ProcNo=0); // Constructor
    int Run(void);      // Run method
    double PIDCalc(void);     // Do the PID calculation
 };
// Define Global variable indexes
#define PID_PROCESS 0
#define PID_ACTUATION 1
#define PID_ERROR 2
#define PID_SET 3

// Define message boxes for data input to PID Control
// The derived task should not have any message boxes since it has
// no run function but, just in case, these are defined at the top
// of the message box set!
#define PID_START NUM_MESSAGE_BOX-1
#define PID_SETPOINT NUM_MESSAGE_BOX-2
#define PID_KP NUM_MESSAGE_BOX-3
#define PID_KI NUM_MESSAGE_BOX-4
#define PID_KD NUM_MESSAGE_BOX-5
#define PID_MINV NUM_MESSAGE_BOX-6
#define PID_MAXV NUM_MESSAGE_BOX-7
#define PID_NEWGAINS NUM_MESSAGE_BOX-8
#define PID_STOP NUM_MESSAGE_BOX-9
```

The values for the task pointers (for process value and actuation) and the associated message box or global variable index are used in the Run() function (of pid_gen1.cpp) to access this information:

```
val = GetGlobalData(ProcessTask,ProcValIndex);
```

to get the process value, and,

```
if(ActuationTask != NULL)
    {
    // Send out the actuation value
    SendMessage(ActuationTask,ActuationBox,0,mc,MESSAGE_OVERWRITE);
    }
```

to send out a new actuation value. If the pointer to the actuation task is NULL no information is sent out but the actuation value is still available as a global value of the control task.

The simulation code for the belt axes and the ovens must be changed to reflect this method of transmitting the actuation information. The relevant change is in the Run() function, shown here for the belt axis simulation,

```
if(GetMessage(BELTAXISMSG_Cur,&MsgFlag,&val))cur = val;
```

Once the PID control classes and their associated member functions are written, PID control tasks can be set up without writing any code at all. In this case, there are four such tasks, two for the belt axes and two for the ovens. This is shown in GlueMain.cpp for one of the belt axes. The others are similar.

```
BeltControl1 = new PIDControl("BeltControl1",0.1, // sample time
        BeltAxis1,BELTAXISMSG_Cur, // where to send actuation
        BeltAxis1,BELTAXIS_x, // where to get process value
        6.0,0.0,10.0,-5.0,5.0, // gains
        1.2,1, // Setpoint, start with controller ON
        PROCESS_A);
```

The constructor is set up with a sufficient parameter coded so that the control can be run independently, that is, there is no requirement for another task to send it set points, turn it on, and so on. This is very useful for this step in the project because the control can then be tested and tuned in a simple software environment. The controller, however, does include code using message boxes to send it new setpoints and/or new gains.

15.6.3 Results

Figures 15-4 and 15-5 show the responses to a constant setpoint for the oven and the belt, respectively. In each case, different setpoints are used showing that the control tasks really are independent even though no additional code was written for them.

15.7 Better Control of Motion (Glue03)

Positioning the belt using the PID controller does not give sufficient control of the smoothness of the motion. This is because the rate at which the motor torque will change depends on how far the belt is from its desired (setpoint) position. If it is far away, there will be a large, sudden change in the command to the motor along with the attendant high acceleration of the object. Smooth motion is required for this system because of the weak glue joint that holds the cylinder to its base during the motion from the assembly area to the oven where the glue is cured. To control motion smoothness a motion profile is applied. When this is done changes in motor excitation are limited to predetermined amounts, regardless of how long a move is requested.

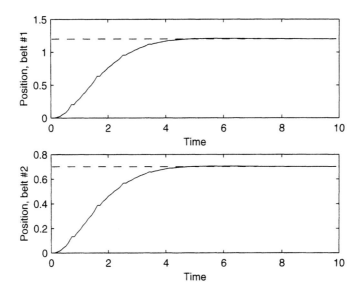

Figure 15-4. PID control of belt axis

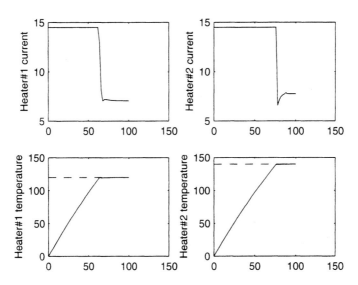

Figure 15-5. PID control of oven temperature

15.7.1 Trapezoidal Motion Profile

Feedback controllers work best when the error (difference between setpoint and measured value of the process variable) is kept small. This can be done by always giving the controller a realistic setpoint. A large, quick change in setpoint is not realistic in the sense that the dynamics of the system prevent it from making a move that quickly so there will be a large error for a long time. During this period the controller often saturates and, in general, the behavior of the system is not very predictable. In a profiled move, the setpoint is changed slowly enough so that the controller can realistically keep up, thereby assuring that the error is always small.

The simplest such profile is trapezoidal in velocity—it consists of a period of constant acceleration (linearly changing velocity), followed by a period of constant velocity, ending with a constant deceleration to the target position. Figure 15-6 shows a sample of a trapezoidal profile.

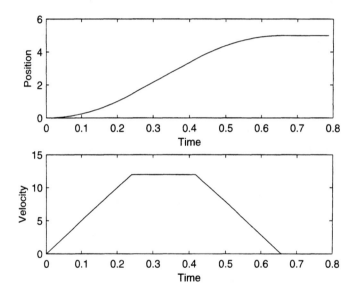

Figure 15-6. Trapezoidal profile

15.7.2 Motion Profile Class

A separate task is used to implement the motion profile. This allows for considerable flexibility in the system design. For example, any kind of controller task can be used, not just PID, as long as it can accept a setpoint input. A debugging version of the program is implemented in this section—it generates a motion profile based on information set in its constructor. This allows for debugging of the basic motion profile generation without the complexity of a command and data passing structure.

As earlier, a generic class is defined from which profile generator tasks can be instantiated. This is the first task in this example that is complex enough to require internal state structure, so the general structure for dealing with states will be demonstrated.

Section 15.7. Better Control of Motion (Glue03)

The motion profile class is quite simple,

```
class CMotion : public BaseTask
    {
    public:
        CMotion(char *name,BaseTask *aProfileOut,
            int aProfileOutBoxNo,double dt,int ProcNo);
        int Run(void);

    private:
        double StopPoint(double xc,double vc,double ac);
        double Sign(double x);
        int Straddle(double x0,double x1,double x2);
        double xtarget,xcur,vcur,acur;
        double accel,vcruise;
        int command;
        BaseTask *ProfileOut;
        int ProfileOutBoxNo;
        int InitialState;    // Mainly for debug
    };
// Global Data base assignments
#define MOTION_xcur 0
#define MOTION_STATE 1
```

A message box, passed through constructor arguments, is used to send the profile output as a position setpoint to an associated controller task. If for some reason this is not convenient, the profile output is also available as a global variable.

15.7.3 Profiler State Structure

The states of the profiler include those to carry it through the different profile zones, one for holding the final position when the profile is done, another to provide for a controlled immediate stop, and yet another for idling when no control is in effect. Figure 15-7 shows the state diagram.

Acceleration and cruise are combined into a single state by limiting the velocity to the cruise velocity. Several of the transitions depend on commands from other tasks. The command mechanism is not implemented in this version. The initial state is set to Accel so that it will immediately execute a profile.

The code associated with implementing the state diagram is in the Run() function in Motion.cpp. To allow for reference to states by name a set of #define statements is used,

```
// Define states:
#define INITIALIZE 0
#define IDLE 1
#define HOLD 2
#define DECEL 3
#define ACCEL 4
#define STOP 5
```

The numbers used are completely arbitrary so there is no need to worry about keeping the order or skipping numbers when the state diagram is edited.

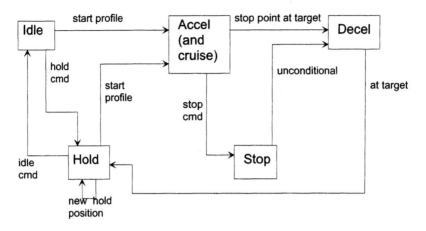

Figure 15-7. State diagram of motion profiler

Prior to doing the state scan, a test is executed to see if the current scan is the first time for a new state,

```
if (NextState != -1)
    {
    // record audit trail here if desired
    AuditTrail(this, State, NextState);
    State = NextState;
    NextState = -1;
    RunEntry = 1;
    }
else
    RunEntry = 0;
```

This sets the flag to run the state's entry function if this is the first scan for a new state. It also sets the NextState variable to -1 as its default value. It will keep this value if no transition tests are true indicating that the next scan should be to the same state.

The state diagram is implemented with a switch statement,

```
switch (State)
    {
    case IDLE:
        // This state will be implemented at the next
        //   stage when a command structure is added
        // Entry
        // No action
        // Test/Exit:

        break;
    case HOLD:
        if(RunEntry)
```

Section 15.7. Better Control of Motion (Glue03)

```
            {
            // Send target position
            SendMessage(ProfileOut,ProfileOutBoxNo,0,xtarget,
                MESSAGE_OVERWRITE);
            }
        // No action
        // Test/Exit
        // This task is terminal for the debugging version
        break;
    case ACCEL:
        {
        // Entry
        if(RunEntry)
            {
            // Compute proper acceleration
            xstop = StopPoint(xcur,vcur,accel);
            if(xstop < xtarget)acur = accel;
            else acur = -accel;
            }
        // Action
        vcur += acur * DeltaTaskTime;
        if(vcur > vcruise)vcur = vcruise;
        if(vcur < (-vcruise))vcur = - vcruise;
        xnext = xcur + vcur * DeltaTaskTime;
        xstop = StopPoint(xnext,vcur,accel);

        // Write result
        xcur = xnext;
        PrivateGlobalData[MOTION_xcur] = xcur;
        if(ProfileOut != NULL)
            {
            // Send result to target task
            SendMessage(ProfileOut,ProfileOutBoxNo,
                0,xcur,MESSAGE_OVERWRITE);
            }
        // Test/exit
        // Check to see if acceleration should be reversed
        if(acur > 0)
            {
            // Only check if velocity and accel are same sign
            if(xstop >= xtarget)
                NextState = DECEL;
            }
        else
            {
            if(xstop <= xtarget)
                NextState = DECEL;
            }
    break;
    }

    case DECEL:
        {
        // Entry
        if(RunEntry)
```

```
            {
            // Compute proper acceleration
            acur = -Sign(vcur)* accel;
            }
        // Action
        // Adjust acceleration to hit target point as
        //   closely as possible
        double aadj = -(vcur * vcur) / (2.0 * (xtarget - xcur));
        vcur += acur * DeltaTaskTime;
        // vcur += aadj * DeltaTaskTime;
        if(vcur > vcruise)vcur = vcruise;
        if(vcur < (-vcruise))vcur = - vcruise;
        xnext = xcur + vcur * DeltaTaskTime;
        xstop = StopPoint(xnext,vcur,accel);

        // Write result
        xold = xcur;
        xcur = xnext;
        PrivateGlobalData[MOTION_xcur] = xcur;
        if(ProfileOut != NULL)
            {
            // Send result to target task
            SendMessage(ProfileOut,ProfileOutBoxNo,
                0,xcur,MESSAGE_OVERWRITE);
            }
        // Test/Exit
        // Check for reaching the target point
        if(((acur * vcur) >= 0)   // Velocity has changed sign
            || Straddle(xold,xtarget,xnext))  // Position at target
            {
            // Profile is complete -- send message
            NextState = HOLD;
            done = 0;  // If this is an intermittent task, run
                // HOLD immediately
            }
        break;
        }
case STOP:
        // This state will need a command structure for
        //   it to be reached
        // Entry
        if(RunEntry)
            {
            xtarget = StopPoint(xcur,vcur,accel);
                // Stop as soon as possible
            }
        // No action
        // Test/Exit
        NextState = DECEL;   // Unconditional transition

        acur = -Sign(vcur) * accel;
    break;
default:
    StopControl("<Motion> Illegal state");
}
```

Each state is a case. Within the case the sections for entry, action, and test/exit are clearly marked.

The function `StopPoint()` is used to compute where the profile would stop if it were decelerated from the position and velocity given in its argument using the acceleration given as the third argument. This is used to figure out when to switch to the DECEL state and is also used in the STOP state to compute a new target position.

15.7.4 Round-Off Error

The object of the motion calculation is that at the end of the profile the velocity reaches zero simultaneously with the position reaching its target. Because the motion task can only be executed at discrete moments in time (as is true with all digital calculations) errors in the calculation will cause the simultaneity condition on which the motion profile is predicated to be violated. That is, if the profile is deemed to end when the velocity reaches zero, the position may well not be at its target. If the position at target is used to determine the end, there may be an error in velocity.

This dilemma is resolved by relaxing the acceleration requirement during the deceleration phase of the profile. At each step, the acceleration is recomputed in such a way that the position and velocity will reach their respective targets simultaneously. The code to do that is in the DECEL state and is repeated here:

```
// Adjust acceleration to hit target point as
//   closely as possible
double aadj = -(vcur * vcur) / (2.0 * (xtarget - xcur));
```

Either the adjusted acceleration, `aadj`, or the specified acceleration, `acur`, can be used to check the size of the error (switch back and forth by commenting out the appropriate statement). The results shown in figure 15-8 use the specified acceleration, as can be seen in the example code. Using the adjusted acceleration makes no discernible difference in this case, but could in cases using a larger step size relative to the profile length.

15.7.5 Discretization Errors in Simulation

Numerical solution of differential equations is an approximation. The error in the solution is a function of the step size or discretization interval. The rule-of-thumb for setting the step size is that it must be small enough so that the qualitative nature of the solution becomes independent of step size.

This problem has the usual discretization errors to contend with, but it has, in addition, a secondary discretization done in setting the time granularity for the tasks and the fact that the simulation includes both continuous time (differential equation) elements and discrete time (difference equation) elements. The continuous time elements are the belt axes and the ovens. The discrete time elements are the PID controllers.

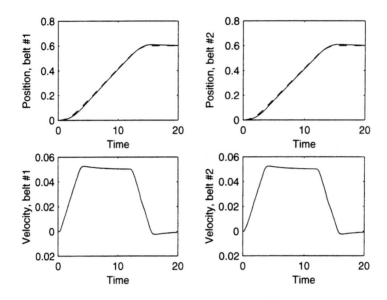

Figure 15-8. Point-to-Point move with trapezoidal profile

The unique element as compared with other simulations is the use of a secondary discretization to set up the time granularity for the task execution. In simulation mode (as are all of the examples thus far) the value TickRate sets the time granularity. Its default value is in the time_mod.hpp file. It can be changed prior to beginning execution of any tasks via the function call to SetTickRate(), which expects the time granularity as its argument (in seconds).

Figure 15-8 was done using a TickRate of 5 x 10 - 5 sec. When the time granularity is increased to 4 x 10 - 4 sec. (figure 15-9) artifacts begin appearing in the solution. These are probably due to the interaction of the simulation and control tasks, which is affected by the TickRate. The size of the artifacts stems from the strong D (derivative) action in the controller.

One way to alleviate this problem is to run the critical tasks more often than other tasks. This can be accomplished by putting the simulation and control tasks into the Intermittent list instead of the LowPriority list. Tasks in the Intermittent list run after every scan of each task in the LowPriority list. Figure 15-10 shows the result of putting the simulation and control tasks for axis #1 in the Intermittent list, but not that of axis #2. The artifacts disappear for axis #1, but remain for axis #2. Another possible approach to this problem would be to use time-stamped data for the controller's process variable so a more accurate derivative can be calculated.

Section 15.7. Better Control of Motion (Glue03)

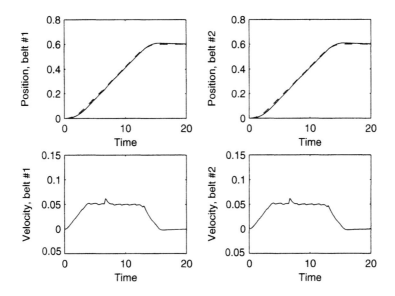

Figure 15-9. Trapezoidal profile, tickrate = 4 x 10 - 4 sec

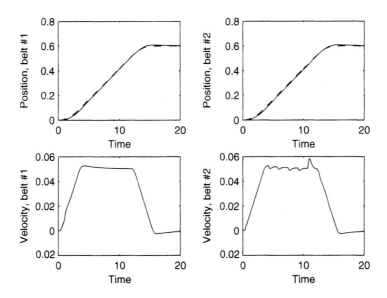

Figure 15-10. Trapezoidal profile: axis #1 tasks in intermittent list, tickrate = 1.0 x 10 - 4

15.8 A Command Structure for Profiled Motion (Glue04)

In order to be useful, the motion task needs to have a command structure so that other tasks can move the axes as needed. Unlike the simple on/off commands used by the PID control task, however, completion of a commanded action takes a significant amount of time. Before proceeding, the requesting task must know whether the action has been completed. A number of ways of doing this come to mind, including monitoring the relevant global variables to see when the action has completed, but the cleanest solution seems to be one in which the task performing the requested action actively signals when the action is complete. This allows for changes in how the end of the action is detected, for example, without changing any of the requesting tasks.

15.8.1 Message-Based Command Structure

Messages have already been used as a convenient medium for commands because they are cleared on being read so there is never any ambiguity as to how many commands have been issued. In this example that usage is extended to include a return message when the action is complete. This is done by including a return message box as the integer value in the message.

A new set of tasks is created here to exercise the motion tasks. These tasks, Transport1 and Transport2, will ultimately be used to control the movement of assembly object from the assembly station to the oven and then to the unload station. In this and the next couple of examples, however, they are just used to test the operation of the control tasks for various system components.

In this and the next example the two transport tasks are constructed independently, rather than using the common-class structure that has been used thus far. This is done to illustrate the construction of independent tasks where the class definition supports only a single state. For the example physical system, this might happen if the second arm were added to the system after the original was built, and was not physically identical. Since the two arms are the same, the classes will be re-united in a subsequent section.

The usage of the command/completion structure is shown from an example in the Run() function of CTransport1, in file trnsport.cpp,

```
// Send a move command to belt #1 Motion task
SendMessage(BeltMotion1,MOTION_CMD_START_PROFILE,TRANSPORT1_MOVE_DONE,1.0);
```

In this message, TRANSPORT1_MOVE_DONE is the message box in the requesting task to which the motion task will send completion information. The associated transition test, which causes a state transition when the move is complete, is,

```
// Check for completion of the move
if(GetMessage(TRANSPORT1_MOVE_DONE,&MsgIndex,&MsgValue))
    {
    // Move is finished, wait for axis to settle
    NextState = WAIT1;
    }
```

Section 15.8. A Command Structure for Profiled Motion (Glue04)   261

The code that receives the command is in the motion task, motion.cpp. The code to receive this command is in the transition/test section of the relevant states,

```
else if(GetMessage(MOTION_CMD_START_PROFILE,&MsgIndex,&MsgVal,&FromTask))
    {
    xtarget = MsgVal;    // Start a new point-to-point move
    fromStart = FromTask;
    respStart = MsgIndex;
    NextState = ACCEL;
```

It saves the information on which task and message box to respond to. The actual response occurs when the move is complete (in the DECEL state),

```
// Profile is complete -- send message
SendCompletionMessage(fromStart,respStart,MOTION_SUCCESS);
```

15.8.2 State Transition Audit Trail

States in each task are specified by numbers so the switch structure can process them. For internal readability it is convenient to use #define statements to give names to the states. If the AuditTrail() function is called when the task is initialized, the state transitions for the task will be saved for printing when the program finishes. By default, the states will be identified by number in the audit trail file. This is hard to read, however, so a provision is available to connect names to the states as printed in the audit trail. This is done by registering the state names in the task's constructor, shown here for the Transport1 task,

```
CreateStateNameArray(20);    // So the audit trail will have
// state names
RegisterStateName("Initialize",0);
RegisterStateName("Start",1);
RegisterStateName("Move1",2);
RegisterStateName("Quit",3);
RegisterStateName("Wait1",4);
RegisterStateName("Move2",5);
RegisterStateName("Wait2",6);
```

The audit trail for this program is (trail.txt):

```
 0.00065 BeltMotion1 IDLE IDLE
 0.0007 BeltMotion2 IDLE IDLE
 0.00095 Transport1 Start Start
 0.001 Transport2 Start Start
 0.0016 Transport1 Start Move1
 0.00165 Transport2 Start Move1
 0.4004 BeltMotion1 IDLE ACCEL
 0.40045 BeltMotion2 IDLE ACCEL
 6.2002 BeltMotion2 ACCEL DECEL
12.0005 BeltMotion1 ACCEL DECEL
14.0013 Transport2 Move1 Wait1
14.2007 BeltMotion2 DECEL HOLD
16.5026 Transport2 Wait1 Move2
```

```
16.8001 BeltMotion2 HOLD ACCEL
22.6006 BeltMotion2 ACCEL DECEL
24.2013 Transport1 Move1 Wait1
24.4006 BeltMotion1 DECEL HOLD
26.7025 Transport1 Wait1 Move2
27 BeltMotion1 HOLD ACCEL
30.401 Transport2 Move2 Wait2
30.6004 BeltMotion2 DECEL HOLD
32.9023 Transport2 Wait2 Quit
36.0001 BeltMotion1 ACCEL DECEL
44.8015 Transport1 Move2 Wait2
45.0002 BeltMotion1 DECEL HOLD
47.3027 Transport1 Wait2 Quit
```

Each line is a state transition event. The first item is the time of the event, the second is the name of the task, the third is the state from which the transition is taking place, and the last item is the state to which the transition goes.

The sequence of events can be extracted from this by following particular trails. Axis 1, for example, starts a move at t = 0.0016s (in task Transport1), completes the move and starts a wait at t = 24.2013s, completes the wait and starts a second move at t=26.7025s, completes that move and starts another wait at t=44.8015s, and, finally, completes the second wait and causes the program to quit at t=47.3027. The motion states needed to implement all of this can be traced in a similar fashion.

This can be displayed in a compact, graphical format by using the same information in numeric format (which is in file `trl_num.txt`). A Matlab program, PlotStates.m, is used to put this information into a graphical time sequence, figure 15-11. Task numbers are listed on the y-axis and time is on the x-axis. For each task, each state # it enters is listed going across, at the proper time. This graph gives a good feeling for the activity but is not a complete record because overprinting causes some of the states to be illegible.

The state numbers are given above, and the task numbers can be found in the file scans.txt, which also lists the number of scans each task gets,

```
Task #   Task Name      #Scans

0        Data Logger    72736
1        BeltAxis1      72736
2        BeltAxis2      72736
3        ThOven1        72736
4        ThOven2        72736
5        BeltControl1   72736
6        BeltControl2   72736
7        HeatControl1   72736
8        HeatControl2   72736
9        BeltMotion1    72736
10       BeltMotion2    72736
11       Transport1     72736
12       Transport2     72736
```

Following that same sequence, Transport1 is task #11 and its state sequence can be read from the graph. The graphical view gives a quick overview of state changes

Section 15.9. Clamps (Glue05)

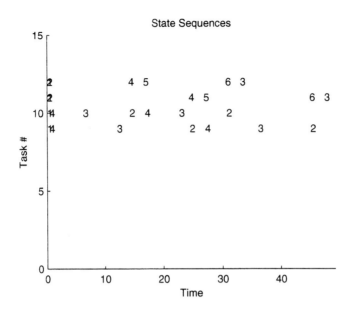

Figure 15-11. Time sequence of states

in one state relative to others although if state changes are too close in time they could come out unreadable.

15.8.3 Motion Results

Figure 15-12 shows the simulated motion of the two axes. The two transport tasks have the same state structure, but use different parameters to achieve the motions shown. The results also show that the profile parameters, acceleration, etc., can be varied so the delicacy of the motion can be changed depending on whether the glue has been cured or not.

15.9 Clamps (Glue05)

Each belt has a clamp fixed to it that is used to attach the assembly object to the belt so it can be moved. The simulation for the clamp operation is quite simple: a command is received to open or close the clamp, there is a delay for the operation, and then the clamp is at its new position.

The clamp tasks are included with the other simulations in sim.cpp. The tasks for the two clamps are instantiated from the same class, CClamp. Its states are:

```
#define CLAMP_INITIALIZE 0
#define CLAMP_OPEN 1
#define CLAMP_CLOSED 2
```

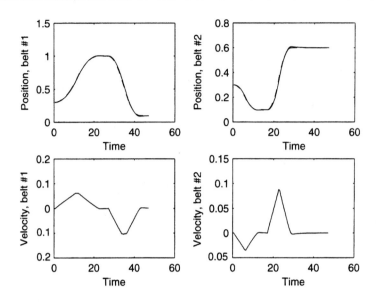

Figure 15-12. Multiple belt axis moves

```
#define CLAMP_MOVINGTO_OPEN 3
#define CLAMP_MOVINGTO_CLOSED 4
```

A global variable, CLAMP_STATUS, is used to monitor where the clamp is by the requesting task. Again, the transport tasks are used to exercise the clamp operation. The state transitions for test operation are:

```
0.00075    BeltMotion1  IDLE IDLE
0.0008     BeltMotion2  IDLE IDLE

0.00105    Transport1   Start Start
0.0011     Transport2   Start Start
0.00135    Clamp1       CLAMP_OPEN CLAMP_OPEN
0.0014     Clamp2       CLAMP_OPEN CLAMP_OPEN
0.0018     Transport1   Start Close1
0.00185    Transport2   Start Close1
0.00285    Clamp1       CLAMP_OPEN CLAMP_MOVINGTO_CLOSED
0.0029     Clamp2       CLAMP_OPEN CLAMP_MOVINGTO_CLOSED
0.50395    Clamp1       CLAMP_MOVINGTO_CLOSED CLAMP_CLOSED
0.504      Clamp2       CLAMP_MOVINGTO_CLOSED CLAMP_CLOSED
0.50515    Transport1   Close1 Wait1
0.5052     Transport2   Close1 Wait1
2.00595    Transport2   Wait1 Open1
2.007      Clamp2       CLAMP_CLOSED CLAMP_MOVINGTO_OPEN
2.5083     Clamp2       CLAMP_MOVINGTO_OPEN CLAMP_OPEN
2.5095     Transport2   Open1 Wait2
3.0062     Transport1   Wait1 Open1
```

```
3.00725    Clamp1       CLAMP_CLOSED CLAMP_MOVINGTO_OPEN
3.3109     Transport2   Wait2 Quit
3.5085     Clamp1       CLAMP_MOVINGTO_OPEN CLAMP_OPEN
3.5097     Transport1   Open1 Wait2
5.11105    Transport1   Wait2 Quit
```

This shows the sequence of commands from the transport tasks and the responses of the clamp tasks. For example, at t=2.00595s Transport2 entered its Open1 state, which sends a command to Clamp2 to open. At t=2.007s Clamp2 started the process of opening. The process completed at t=2.5083s when Clamp2 entered its state CLAMP_OPEN. Transport2 then went to state Wait2 at t=2.5095s.

15.10 Robots (Glue06)

The robots are simulated in basically the same way as the clamps. There are more states, but the idea is the same: a command causes entry to a state the is a time delay while the robot goes to its commanded position. At present, there are no underlying physical models for any of the robots. Such models could be added without affecting this high level interface at all.

There are four robot-like devices in this system: an assembly robot to place the base and cylinder pieces on the assembly platform, a glue applicator, and two unload robots, one for each arm. Each of these is represented as a simulation task. The unload robots have only a single class; each of the others has its own class. These are all placed with the simulations. Depending on how an actual robot were to be implemented (or a more detailed simulation) the present tasks could become either tasks that communicate with an external robot controller or tasks that are the top-level for direct robot control.

The interface between the robot simulation tasks and tasks that are sending commands is the same as is used for the motion task. That is, the command is sent in the form of a message with a return message box used for the robot to signal completion. Robot state and status are also available as global variables.

The difference between using messages and using global variables for these sorts of command interfaces is important in deciding how to set up controlling and controlled tasks. When a status variable is used there can be the problem of repeated values or missed values. For example, if a controlled system is in a ready status a command can be sent to it to do something. The sending task often needs to know when the activity has been completed. It might try to determine completion by monitoring the status to see when the controlled task returns to ready. If it checks too quickly, however, the controlled task may not have actually started anything yet and still shows that it is ready. The requesting task would then mistakenly think that the action had been completed. Messages avoid this problem because they are cleared when they are read so cannot be reread mistakenly.

A short section of the state transition audit trail for the robot control shows how this operates. Again, the transport tasks are used to exercise the components on each of the axis arms. The independent transport task classes have been replaced with a common class. A new task (assembly.cpp) is added to operate the load (assembly) robot and the glue applicator.

```
...
0.00275    LoadRobot       Start AtReady
0.0028     UnloadRobot1    Start AtReady
0.00285    UnloadRobot2    Start AtReady
0.0029     GlueApplicator  Start AtReady
0.00365    Clamp1          CLAMP_OPEN CLAMP_MOVINGTO_CLOSED
0.0037     Clamp2          CLAMP_OPEN CLAMP_MOVINGTO_CLOSED
0.00375    LoadRobot       AtReady PickupBase
0.50475    Clamp1          CLAMP_MOVINGTO_CLOSED CLAMP_CLOSED
0.5048     Clamp2          CLAMP_MOVINGTO_CLOSED CLAMP_CLOSED
0.5064     Transport1      Close1 Wait1
0.50645    Transport2      Close1 Wait1
1.75545    LoadRobot       PickupBase PlaceBase
1.90815    Transport2      Wait1 Open1
1.9095     Clamp2          CLAMP_CLOSED CLAMP_MOVINGTO_OPEN
2.41075    Clamp2          CLAMP_MOVINGTO_OPEN CLAMP_OPEN
2.4124     Transport2      Open1 Wait2
3.0077     Transport1      Wait1 Open1
3.00905    Clamp1          CLAMP_CLOSED CLAMP_MOVINGTO_OPEN
3.5074     LoadRobot       PlaceBase GoingToReady
3.5103     Clamp1          CLAMP_MOVINGTO_OPEN CLAMP_OPEN
3.51195    Transport1      Open1 Wait2
5.1138     Transport1      Wait2 UnloadAssembly
5.1148     Transport1      UnloadAssembly WaitClear
5.1153     UnloadRobot     AtReady PickupAssembly
6.0087     LoadRobot       GoingToReady AtReady
6.01035    Assembly        PlaceBase ApplyGlue
...
```

15.11 Cure/Unload (Glue07)

All of the components needed to operate the system are now in place. This section covers the sequencing of the assembled object from the assembly area to the oven and then to the unload area. This is accomplished by the transport tasks, which share the CTransport class.

The state logic for this operation is linear, that is, a straight sequence of states that carry out the required operations. The states are:

```
#define Initialize 0
#define Start 1
#define AtReady 2
#define OpenClamp1 8
#define MoveToAssy 3
#define CloseClamp 4
#define MoveToOven 5
#define Cure 6
#define MoveToUnload 7
#define OpenClamp2 11
#define Unload 9
#define MoveToReady 10
```

Section 15.11. Cure/Unload (Glue07)

The states are listed in their order of execution rather than in numerical order. For this simple transition structure, then, this list can stand in for the state transition diagram. The transition conditions for each state are completion of the indicated activity. Some of the transitions use completion messages as tests, such as states that move the belt axis, and others use global status variables, such as the clamp status.

The `Assembly` task is used to trigger the transport activities. It starts the #1 transport task operating, waits for the object to clear the assembly area, then starts the #2 transport task. When both tasks have returned to their `AtReady` status `Assembly` ends the program.

Figure 15-13 shows the motion of the belt axes during this process. Both axes go to their Ready positions on startup. Axis #1 starts executing the curing sequence first. As soon as the object is at the oven position (and therefore clear of the assembly area) the second axis starts.

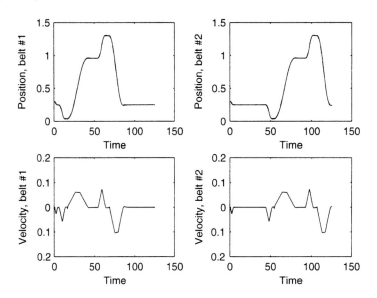

Figure 15-13. Motion of belt axis during curing sequence

Figure 15-14 shows the sequences of states for all of the tasks. The task list (from file `scans.txt`) is:

```
Task #    Task Name     #Scans
0         Data Logger   147525
1         BeltAxis1     147525
2         BeltAxis2     147525
3         ThOven1       147525
4         ThOven2       147525
5         Clamp1        147525
```

```
 6    Clamp2          147525
 7    LoadRobot       147525
 8    UnloadRobot1    147525
 9    UnloadRobot2    147525
10    GlueApplicator  147525
11    BeltControl1    147525
12    BeltControl2    147525
13    HeatControl1    147525
14    HeatControl2    147525
15    BeltMotion1     147525
16    BeltMotion2     147525
17    Transport1      147525
18    Transport2      147525
19    Assembly        147525

Total scans, all tasks: 2950500
Total running time: 147.602
```

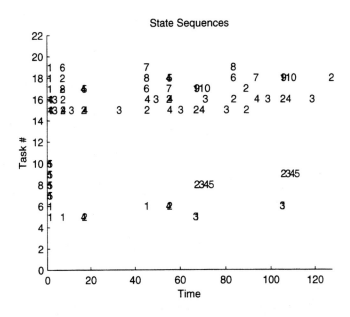

Figure 15-14. State sequences for curing process

The overlapped activity of the two transport tasks (# 17 and #18) can be seen clearly from the figure.

The complete state transition audit trail is shown next. It can be used to do the basic verification that the system is operating properly.

```
0.001     BeltMotion1   IDLE  IDLE
0.00105   BeltMotion2   IDLE  IDLE
0.0013    Transport1    Start Start
0.00135   Transport2    Start Start
```

Section 15.11. Cure/Unload (Glue07)

```
 0.0014    Assembly        Start Start
 0.00165   Clamp1          CLAMP_OPEN CLAMP_OPEN
 0.0017    Clamp2          CLAMP_OPEN CLAMP_OPEN
 0.00175   LoadRobot       Start Start
 0.0018    UnloadRobot1    Start Start
 0.00185   UnloadRobot2    Start Start
 0.0019    GlueApplicator  Start Start
 0.00275   LoadRobot       Start AtReady
 0.0028    UnloadRobot1    Start AtReady
 0.00285   UnloadRobot2    Start AtReady
 0.0029    GlueApplicator  Start AtReady
 0.40005   BeltMotion2     IDLE ACCEL

 0.401     BeltMotion1     IDLE ACCEL
 2.20095   BeltMotion1     ACCEL DECEL
 2.201     BeltMotion2     ACCEL DECEL
 5.80215   Transport1      Start AtReady
 5.8022    Transport2      Start AtReady
 5.80325   Assembly        Start TestTran1
 5.80515   Transport1      AtReady OpenClamp1
 5.80615   Transport1      OpenClamp1 MoveToAssy
 5.8065    Clamp1          CLAMP_OPEN CLAMP_OPEN
 6.00095   BeltMotion1     DECEL HOLD
 6.001     BeltMotion2     DECEL HOLD
 6.20105   BeltMotion1     HOLD ACCEL
10.001     BeltMotion1     ACCEL DECEL
15.4022    Transport1      MoveToAssy CloseClamp
15.4035    Clamp1          CLAMP_OPEN CLAMP_MOVINGTO_CLOSED
15.601     BeltMotion1     DECEL HOLD
15.9048    Clamp1          CLAMP_MOVINGTO_CLOSED CLAMP_CLOSED
15.9064    Transport1      CloseClamp MoveToOven
16.2003    BeltMotion1     HOLD ACCEL
31.6004    BeltMotion1     ACCEL DECEL
43.8021    Transport1      MoveToOven Cure
43.8022    Assembly        TestTran1 TestTran2
43.8042    Transport2      AtReady OpenClamp1
43.8055    Clamp2          CLAMP_OPEN CLAMP_OPEN
44.0009    BeltMotion1     DECEL HOLD
53.8034    Transport1      Cure MoveToUnload
54.2003    BeltMotion1     HOLD ACCEL
59.0008    BeltMotion1     ACCEL DECEL
65.6016    Transport1      MoveToUnload OpenClamp2
65.6029    Clamp1          CLAMP_CLOSED CLAMP_MOVINGTO_OPEN
65.8004    BeltMotion1     DECEL HOLD
66.1042    Clamp1          CLAMP_MOVINGTO_OPEN CLAMP_OPEN
66.1058    Transport1      OpenClamp2 Unload
66.1059    Transport2      OpenClamp1 MoveToAssy
66.1073    UnloadRobot1    AtReady PickupAssembly
66.4007    BeltMotion2     HOLD ACCEL
68.1584    UnloadRobot1    PickupAssembly PlaceAssembly
68.1599    Transport1      Unload MoveToReady
68.4007    BeltMotion1     HOLD ACCEL
70.2007    BeltMotion2     ACCEL DECEL
70.2095    UnloadRobot1    PlaceAssembly GoingToReady
72.4106    UnloadRobot1    GoingToReady AtReady
```

```
75.802    Transport2      MoveToAssy CloseClamp
75.8033   Clamp2          CLAMP_OPEN CLAMP_MOVINGTO_CLOSED
76.0008   BeltMotion2     DECEL HOLD
76.3046   Clamp2          CLAMP_MOVINGTO_CLOSED CLAMP_CLOSED
76.3062   Transport2      CloseClamp MoveToOven
76.6001   BeltMotion2     HOLD ACCEL
79.0003   BeltMotion1     ACCEL DECEL
87.6021   Transport1      MoveToReady AtReady
87.8009   BeltMotion1     DECEL HOLD
92.0002   BeltMotion2     ACCEL DECEL
104.202   Transport2      MoveToOven Cure

104.202   Assembly        TestTran2 WaitTran
104.401   BeltMotion2     DECEL HOLD
114.203   Transport2      Cure MoveToUnload
114.6     BeltMotion2     HOLD ACCEL
119.401   BeltMotion2     ACCEL DECEL
126.002   Transport2      MoveToUnload OpenClamp2
126.003   Transport2      OpenClamp2 Unload
126.004   Clamp2          CLAMP_CLOSED CLAMP_MOVINGTO_OPEN
126.005   UnloadRobot2    AtReady PickupAssembly
126.2     BeltMotion2     DECEL HOLD
126.505   Clamp2          CLAMP_MOVINGTO_OPEN CLAMP_OPEN
128.056   UnloadRobot2    PickupAssembly PlaceAssembly
128.057   Transport2      Unload MoveToReady
128.4     BeltMotion2     HOLD ACCEL
130.107   UnloadRobot2    PlaceAssembly GoingToReady
132.308   UnloadRobot2    GoingToReady AtReady
139.001   BeltMotion2     ACCEL DECEL
147.602   Transport2      MoveToReady AtReady
```

Careful examination of this list along with the state transition specification of each of the tasks is a critical verification and first-step debugging tool. In this case, for example, the following sequence fragment seems incorrect:

```
79.0003   BeltMotion1     ACCEL DECEL
87.6021   Transport1      MoveToReady AtReady
87.8009   BeltMotion1     DECEL HOLD
```

The Transport1 task is supposed to move to its AtReady task when the belt motion is done. However, this fragment shows the Transport1 task moving to AtReady before the belt motion is completed (as indicated by its transition to the HOLD state). Detailed examination of the system operation, however, shows that this is not really an error. The sequence as shown occurs in this order because of two factors:

1. Transitions to new states are recorded in the audit trail when the new state is executed for the first time—not when the previous state is executed for the last time.

2. The transport tasks execute on every available scan whereas the motion tasks execute every 0.2 seconds.

Putting these together explains the seemingly incorrect sequence. On reaching the end of a move the motion task sends a completion message as part of the test/exit sequence of the DECEL task. Because is executes on every scan, the transport task sees the information that the move is done immediately and causes a transition to its AtReady state (at $t = 87.6021$s). 0.2 seconds later the motion task executes next (at $t = 87.8009$s) and records its transition to HOLD.

Transition sequences for individual tasks can be broken out by using a spreadsheet (Excel in this case). Using the spreadsheet filtering capability, here are the transitions for the Assembly task,

```
0.0014     Assembly        Start Start
5.80325    Assembly        Start TestTran1
43.8022    Assembly        TestTran1 TestTran2
104.202    Assembly        TestTran2 WaitTran
```

and those for the Transport1 task:

```
0.0013     Transport1      Start Start
5.80215    Transport1      Start AtReady
5.80515    Transport1      AtReady OpenClamp1
5.80615    Transport1      OpenClamp1 MoveToAssy
15.4022    Transport1      MoveToAssy CloseClamp
15.9064    Transport1      CloseClamp MoveToOven
43.8021    Transport1      MoveToOven Cure
53.8034    Transport1      Cure MoveToUnload
65.6016    Transport1      MoveToUnload OpenClamp2
66.1058    Transport1      OpenClamp2 Unload
68.1599    Transport1      Unload MoveToReady
87.6021    Transport1      MoveToReady AtReady
```

These allow close examination of individual task sequences.

15.12 Making Widgets (Glue08)

Putting all of this together to actually manufacture assemblies is almost anticlimactic! The Assembly task must get the proper state sequence to assemble objects. Like the Transport tasks the state diagram is basically linear. The only branch point is in BeginNewObject which will either start building a new object or quit if enough objects have already been built.

The state sequence for Assembly, extracted from the audit trail is shown next. Sequences for the other tasks are almost the same as was shown in the previous section. This sequence makes four objects. The order of states can be seen from the audit trail listing. The Shutdown state waits for both belt axes to reach their Ready positions before stopping. This guarantees that the last object will be completed before stopping the control program.

```
0.0014     Assembly        Start Start
5.80325    Assembly        Start BeginNewObject
5.80425    Assembly        BeginNewObject PlaceBase
```

```
 11.8124    Assembly    PlaceBase ApplyGlue
 29.2206    Assembly    ApplyGlue PlaceCylinder
 34.8285    Assembly    PlaceCylinder WaitForTransport
 34.8295    Assembly    WaitForTransport CureUnload
 73.0016    Assembly    CureUnload BeginNewObject
 73.0026    Assembly    BeginNewObject PlaceBase
 79.0107    Assembly    PlaceBase ApplyGlue
 96.4188    Assembly    ApplyGlue PlaceCylinder
102.027     Assembly    PlaceCylinder WaitForTransport
102.028     Assembly    WaitForTransport CureUnload
140.202     Assembly    CureUnload BeginNewObject
140.203     Assembly    BeginNewObject PlaceBase

146.211     Assembly    PlaceBase ApplyGlue
163.619     Assembly    ApplyGlue PlaceCylinder
169.227     Assembly    PlaceCylinder WaitForTransport
169.228     Assembly    WaitForTransport CureUnload
207.402     Assembly    CureUnload BeginNewObject
207.404     Assembly    BeginNewObject PlaceBase
213.411     Assembly    PlaceBase ApplyGlue
230.818     Assembly    ApplyGlue PlaceCylinder
236.427     Assembly    PlaceCylinder WaitForTransport
236.428     Assembly    WaitForTransport CureUnload
274.603     Assembly    CureUnload BeginNewObject
274.604     Assembly    BeginNewObject Shutdown
```

Once the details of the sequences have been checked and verified, this is a completed simulation. However, it just represents the beginning of the design and implementation work. The simulation can be used to look for ways to optimize production, design of the operator interface, error handling, and so on. It can also be used to check on timing issues by running it as a real-time simulation. As the actual machine is built, the simulation must be transferred to the target environment. At that point, various assumptions that were made in building the simulation can be checked and the simulation can be updated to correspond more closely with the real system. This can be very valuable for future machine modifications or upgrades, for troubleshooting as the machine is initially brought up, and to help debug problems arising in the field.

Chapter 16

THE GLUING CELL EXERCISE IN TRANRUN4

Flexible manufacturing cells are becoming popular for the fabrication of small quantities of high-value equipment. Although they come in different configurations, most are characterized by the use of mechatronic techniques to provide the ability to modify or reconfigure the operation of the cell by changing only the software. The gluing cell project demonstrates the use of the programming techniques taught in the class through the step-by-step development of a program to control the gluing machinery. The C++ language is used to write the control software, and the TranRun4 package is used to implement the task/state and scheduling structure of the program.

16.1 The Gluing System

The purpose of the gluing cell is to glue two parts together with a thermosetting adhesive. The adhesive is applied to the first part (the base) in a gluing machine which applies the glue by moving a glue nozzle in a predetermined path across the object as a glue pump applies the glue. Then the second part is placed on top of the first by a robot. The robot then picks up the parts and places them onto one of two conveyors (figure 16-1). Each conveyor carries the glued assembly into an oven whose temperature is held at the optimal curing temperature of the thermosetting glue. The assembly is kept inside the oven for a predetermined time until the adhesive has cured; then the conveyor moves the finished assembly out of the oven and away from the gluing machine.

Because the heat curing operation takes a longer time to complete than the application of the glue, two independent conveyors are provided to carry parts into the oven. The gluing cell will therefore be able to apply glue to one part while another is curing in the oven.

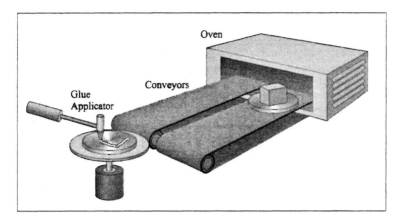

Figure 16-1. Gluing cell layout

16.2 Simulation and Prototyping

A series of exercises will be formed around the development of programs which control the gluing cell. Each of the modules which control different parts of the cell can be developed and tested independently. The components will then be integrated into a complete system which can be tested in simulation, reflecting a bottom-up approach to the software development.

The simulation of the gluing cell's subsystems involves the use of the same dynamic system models which are used in classical control theory. The systems of differential equations which result are solved by the use of simple Euler integrators built into the simulation code. Euler integrators are somewhat crude but easily programmed and can yield satisfactorily accurate results if the time steps are sufficiently small. The simulations are programmed as tasks in the task/state transition structure of the program. This technique allows for the simple switching of the program from simulation mode to prototyping on real hardware, as the simulation code is separated from the control code, and the latter is not changed between the two modes.

16.3 The Project Components

In keeping with the bottom-up development cycle discussed earlier, the first parts of the exercise will create simulations of individual components in the gluing cell. The sequence in which the exercises will be presented follows.

Glue00: The conveyor mechanisms are simulated and tested

Glue01: The oven is simulated and tested

Glue02: PID controllers are applied to conveyors and oven

Glue03: An operator interface is added

Glue04: Motion profiling is added to the conveyor control

Glue05: A supervisor task for profiled motion is created

Glue06: Simulations for loading and unloading robots are added

Glue07: Coordination of profiled belt motion and the unloading robots

Glue08: An overall supervisory task is created to manage the gluing process

16.4 Glue00: Conveyor Simulation

The conveyor device which moves the glued parts in and out of the oven is modeled in this stage. The conveyor uses a DC motor to drive a rotary-to-linear mechanical system which could be a rack and pinion, belt drive, or leadscrew. An incremental encoder measures the motion of the conveyor's carriage. This encoder could be a rotary or linear optical encoder or even a magnetic scale. For the purposes of the simulation, the position measurement transducer and the software which reads and interprets its signals are not modeled; instead, the simulation of the belt produces as an output the position of the carriage as measured by a properly operating measurement subsystem. The measurement subsystem will be tested on the prototype conveyor so that before it is integrated into the real system, we can be reasonably sure that the correct position is indeed available.

16.4.1 The Dynamic Model

A simple dynamic model is used to describe the conveyor system. An electronic power amplifier produces a voltage V which is applied to the DC motor. The current i which flows through the motor is determined by this voltage, the resistance R of the motor, and the back-EMF V_b of the motor (inductance is assumed negligible). Back-EMF is caused by the motor acting as a generator when in motion, and it acts to reduce the current flowing through the motor as the motor turns faster. It is calculated as $V_b = K_\tau \cdot \omega$, where K_τ is the back-EMF constant (a characteristic of the motor) and ω is the speed of rotation of the motor. The current flowing through the motor is then

$$i = \frac{V - V_b}{R}$$

The motor produces a torque τ_m which is proportional to the current flowing through it; the constant of proportionality is K_τ, the same as that which relates back-EMF to speed (if consistent units of measurement are chosen). A frictional torque $\tau_f = f \cdot \omega$ opposes the motion of the carriage; b is the coefficient of friction. The net torque which acts to move the system's mechanical inertia J is then

$$\tau = K_\tau i - b\omega$$

The equations of motion are therefore as follows:

$$\frac{d\omega}{dt} = -\left(\frac{K_\tau^2}{JR} + \frac{b}{J}\right)\omega + \left(\frac{K_\tau}{JR}\right)V$$

$$\frac{d\theta}{dt} = \omega$$

A conveyor system has both linear and angular motion components, and each of these has friction and inertia. In these equations, the linear inertia and friction are lumped together into the angular inertia and friction, thereby modeling the total system which is seen by the motor. The angular position θ and velocity ω are linearly related to linear position x and velocity v by a characteristic radius in the pulleys or leadscrew: $x = r\theta$ and $v = r\omega$.

The Euler integration approximates the right-hand side of each differential equation as constant over each time step and equal to the value of the equation as calculated at the beginning of the time period. Given a pair of coupled differential equations of the form

$$\frac{dx_1}{dt} = f_1(x_1(t), x_2(t), u(t))$$

$$\frac{dx_2}{dt} = f_2(x_1(t), x_2(t), u(t))$$

and given initial conditions for all dependent variables, each step of the Euler integration can be calculated as follows:

$$x_1(t + \Delta t) = f_1(x_1(t), x_2(t), u(t))\Delta t + x_1(t)$$
$$x_2(t + \Delta t) = f_2(x_1(t), x_2(t), u(t))\Delta t + x_2(t)$$

These equations can be slightly modified by noticing that the value of $x_1(t + \Delta t)$ is known before x_2 is computed. The following form, although appearing somewhat more complicated, can be simpler to compute and mathematically more stable than the original form.

$$x_1(t + \Delta t) = f_1(x_1(t), x_2(t), u(t))\Delta t + x_1(t)$$
$$x_2(t + \Delta t) = f_2(x_1(t + \Delta t), x_2(t), u(t))\Delta t + x_2(t)$$

For the conveyor system, the equations of motion and the Euler integration are coded as follows:

```
// Calculate constants A and B once, before simulation is run
A = -((TorqueConst * TorqueConst / (Inertia * Resistance))
    + (FricCoeff / Inertia));
B = (TorqueConst / (Inertia * Resistance));

// Integrations performed at each time step
omega += ((A * omega) + (B * Voltage)) * DeltaTime;
angle += omega * DeltaTime;
position = angle * CharRadius;
```

16.4.2 Creating the Conveyor Task

The relatively simple dynamic equations which describe the conveyor are embodied in a mere five lines of C++ code. However, it takes a great deal more code than that to ensure that these five lines are run at the correct time and that the data needed to perform the simulation is passed to and from the other tasks in the system. Here we discuss in some detail the process by which the conveyor task is built into a task object for use in a TranRun4 program.

The Glue00 program contains two conveyor belt tasks, one for each conveyor in the system. It also incorporates another task whose purpose is to record the motions of the belts in a data file. The data logging task will be discussed in section 16.4.3 below.

The definition of the conveyor belt task is in the file BeltAxis.hpp. Tasks and states are represented in TranRun4 by objects of classes CTask and CState and their descendents respectively. As descendents of CTask and CState, the user's task and state classes implement the algorithms which perform the desired control and simulation functions while inheriting all the functionality of the base task and state classes. The functionality inherited from CTask includes the ability to manage lists of states, perform real-time scheduling, and transfer data between tasks as described in section 16.4.4. From CState is inherited the ability to implement state transitions. The TranRun4 system also supplies a number of useful utilities which profile functions, document the system organization, and create a record of state transitions for later analysis.

In this example problem, the task object is instantiated directly from the standard task class without deriving a user-defined class. This is possible because all of the unique code and data which are added to the conveyor task are stored within the conveyor state object. When many states which belong to the same task need to access data belonging to the task, it may be necessary to derive a task class which holds that data; this is done in the Glue04 module described in section 16.8.

The belt axis task simulates a device which is always running and whose dynamic equations do not change with time. Therefore, the state transition logic representation of this task is very simple: there is just one state. A naming convention has been adopted for TranRun4 programs in which state names are constructed by concatenating a description of the state to the name of the parent task. Thus the state belonging to a task of class C_BeltAxis which represents that axis in a "running" state is called C_BeltAxis_Running. This one state object contains all the user-written code for this task. The user's code is placed in a class which is descended from the standard state class CState. The definition of the belt axis state is shown here.

```
class C_BeltAxis_Running : public CState
    {
    private:
        double Position;          // Current position of the conveyed item
        double Velocity;          // Velocity at which the load is moving
        double Voltage;           // Electrical voltage to the motor
```

```
        double Inertia;              // Inertia of the system as seen by motor
        double TorqueConst;          // Torque and back EMF constant of motor
        double FricCoeff;            // Friction coefficient of the bearings
        double Resistance;           // Electrical resistance of the motor
        double coeff_A;              // Coefficients used in first-order
        double coeff_B;              // dynamic model of motor and load
        double omega;                // Angular speed and angular position
        double theta;                // of the simulated motor shaft
        double CharRadius;           // Translates rotational to linear motion
        real_time PreviousTime;      // Save time values for Euler integrator
    public:
        // The constructor sets default parameters and parent task of this state
        C_BeltAxis_Running (CTask*, char*, double, double, double, double, double);

        void Entry (void);           // We'll overload the Entry() and
        void Action (void);          // Action() functions of class CTask
```

The state object's member data fields—`Position`, `Velocity` and so on—contain all the data which belongs to this task. If there were more states which needed to access the same data, the data could be kept as state member data and accessed by means of state pointers or through function calls; or otherwise the data could be declared as `static` variables placed in the `.cpp` file above all the function definitions. (These methods will be demonstrated in subsequent sections of this chapter.) However, communicating data through the use of static, file-scope global variables or function calls is only acceptable between states *within* a task, and not for the communication of information *between* tasks. Intertask data communication is discussed in detail in section 16.4.4.

The constructor function `C_BeltAxis_Running` is called from the main program to create the belt axis state object before the scheduler is started. It is used not only to instantiate the state object but also to set those parameters which might vary between different instantiations of the object. In the glue exercise, we have two conveyor belts which may or may not be identical pieces of machinery. For example, one conveyor might be an older model with a less powerful motor. Any parameters which can vary in this manner are easily specified in the constructor's parameter list when the constructor is called from within the main program. The constructor's code from `BeltAxis.cpp` is shown here:

```
//------------------------------------------------------------------------------
//  Constructor:  C_BeltAxis_Running
//     The constructor sets starting parameters and parent task of this state.

C_BeltAxis_Running::C_BeltAxis_Running (CTask* pTask, char* aName, double aInert,
                         double aKt, double aFric, double aRes, double aRad)
    : CState (pTask, aName)
    {
    Inertia = aInert;                  // Save member data from parameters
    TorqueConst = aKt;
    FricCoeff = aFric;
    Resistance = aRes;
    CharRadius = aRad;
```

```
    //  Calculate constants A and B once, before simulation is run
    A = -((TorqueConst * TorqueConst / (Inertia * Resistance))
        + (FricCoeff / Inertia));
    coeff_B = (TorqueConst / (Inertia * Resistance));

    CreateDataInbox ("Voltage");              // Create inbox for motor current
    CreateGlobalData ("Velocity", &Velocity); // Create global database items for
    CreateGlobalData ("Position", &Position); //   belt velocity and position
    }
```

Besides copying the parameters which were specified in the constructor's call from the main program, the constructor also calculates coefficients used in the dynamic simulation (this is done just for efficiency of execution) and creates objects which are used to communicate data with other tasks.

In addition to the constructor, the belt axis running state contains two member functions named `Entry()` and `Action()`. These functions contain the code which defines the behavior of the state. A third function, `TransitionTest()`, is programmed by the user to handle the transitions from one state to another; it will be discussed in a subsequent section in which a task with more than one state is constructed.

The `Entry()` function is run once when a state is entered by means of a state transition. For a task with only one state, this means that `Entry()` will run just once when the scheduler is first started. The code of the entry function initializes the variables used in the simulation.

```
//-----------------------------------------------------------------------------
// Function: Entry
//     This function runs once when the C_BeltAxis_Running state is first scheduled.
//     It initializes the integrator's time counter and various other variables.

void C_BeltAxis_Running::Entry (void)
    {
    Position = 0.0;                  // Initialize variables which may be
    Velocity = 0.0;                  // sent to other tasks
    Voltage = 0.0;                   // or received from other tasks
    omega = 0.0;                     // Set initial conditions for
    theta = 0.0;                     // state variables

    PreviousTime = GetTimeNow ();
    }
```

The `Action()` function is run periodically by the task scheduler. During each run, it performs one time step of the belt axis simulation. This involves getting data from other tasks, computing the simulated position and velocity of the conveyor, and making the results available to other tasks. The five lines of code around which this task revolves are to be (finally) seen in this function.

```
//-----------------------------------------------------------------------------
// Function: Action
//     This function runs repeatedly as the simulated belt axis moves. It performs
//     the dynamic system simulation and sends the results to other tasks.
```

```
void C_BeltAxis_Running::Action (void)
    {
    //  Get a new value for the simulated current if one has arrived
    ReadDataPack ("Voltage", NULL, &Voltage, NULL, NULL);

    //  Compute the time elapsed between the last time this function ran and the
    //  current time; this time is used in the Euler integration
    real_time ThisTime = GetTimeNow ();
    real_time DeltaTime = ThisTime - PreviousTime;
    PreviousTime = ThisTime;

    //  Integrations performed at each time step
    omega += ((coeff_A * omega) + (coeff_B * Voltage)) * DeltaTime;
    theta += omega * DeltaTime;
    Position = theta * CharRadius;
    Velocity = omega * CharRadius;

    //  Copy local variables to the global database
    CopyGlobalData ();

    //  This is a sample time task, so it must be idled or it will run all the time
    Idle ();
    }
```

The last line of the action function is a call to `Idle()`. This call informs the scheduler that the task currently executing has finished its duties during the current time period. It is critically important that this function be called here; if it is not, the action function will continue to run again and again, never allowing lower priority tasks to run.

16.4.3   The Data Logging Task

The purpose of the data logging task is to record data from the simulated or real gluing system while the system is running in real time, and to write that data to a file after the real-time run has finished. This is necessary because on most computer systems, writing data to a file can take up a large amount of processor time and often prevents other operations—such as a real-time scheduler—from running during the write operation. This problem is avoided by storing the recorded data in memory while the scheduler is operational, then transferring that data from memory to a disk file after the completion of the real-time run.

A standard data logging class, `CDataLogger`, has been created which automates most of the data handling. This automation permits data to be saved with a minimum of user coding but has the disadvantage that the logged data must follow a fixed format. The data logger object records data in real time and writes the results to arrays which can be easily imported into spreadsheets or mathematical analysis packages such as Matlab. To use the data logger class, one need only instantiate an object of class `CDataLogger` within a task or state's constructor, configure the data logger so that it knows how many columns of what kind of data to record, and then instruct the data logger to save data as the real-time program

Section 16.4. Glue00: Conveyor Simulation

runs. Then deleting the data logger object in the destructor of the logging task (or one of its states) will cause the data to be written to a file.

The data logger task has two states, one for a running data logger (that is, it is currently recording data) and one for a stopped logger. The class definitions for the two state classes follow.

```
//=============================================================================
// State:  C_GlueDataLogger_Running
//     This state saves data while the data logger is active.
//=============================================================================

class C_GlueDataLogger_Running : public CState
    {
    private:
        CDataLogger* TheLogger;          // Generic Data Logger object

    public:
        // The constructor sets name of file to which to write data
        C_GlueDataLogger_Running (CTask* pTask, char* aTaskName, char* aFileName);
        ~C_GlueDataLogger_Running (void);

        void Action (void);              // Action() function with user code
        CState* TransitionTest (void);   // Test for transition to other state
    };

//=============================================================================
// State:  C_GlueDataLogger_Stopped
//     This state doesn't save data, because the logger isn't running now.
//=============================================================================

class C_GlueDataLogger_Stopped : public CState
    {
    public:
        // The constructor sets default parameters and parent task of this state
        C_GlueDataLogger_Stopped (CTask* pTask, char* aName);

        CState* TransitionTest (void);   // Test for transition to other state
    };
```

The data logger running constructor is given the standard parameters for a state object—a pointer to the state's parent task and a name for the state. In addition, there is a file name which is used to specify the name of the file to which logged data will be written. This allows the programmer to create any number of data loggers which simultaneously log different data, then write the data to different files. The states have no `Entry()` functions because there isn't anything which needs to be done once in response to a transition from one state to the other. Only the `C_DataLogger_Running` state has an `Action()` function because only that state has any real work to do. Both states have `TransitionTest()` functions which determine if a condition has occurred in which a transition should occur to the other state.

The constructor and destructor for the running state are shown next. The constructor creates a data logger object and configures it with the following parameters:

- The memory buffer which stores the data is initially allocated to save 1,024 points of data. If this buffer becomes full, another buffer with room for 1,024 more points will be allocated, and the process will continue unless the computer runs out of free memory.

- When the data logger object is deleted in the destructor, the data saved in the memory buffers will be written to a file with the name specified by the parameter aFileName.

- The logger will save data in 5 columns, and all the data will be of type double. Other types of data such as integers can also be saved.

- Default parameters are used for the rest of the logger configuration. For example, the columns in the data file will be delimited by spaces.

```
//-------------------------------------------------------------------------
// Constructor:  C_GlueDataLogger_Running
//     This constructor creates a data logger state, creates the CDataLogger object,
//     and sets name of file to which to write data.

C_GlueDataLogger_Running::C_GlueDataLogger_Running (CTask* pTask, char* aStateName,
                                    char* aFileName) : CState (pTask, aStateName)
    {
    // Create a data logger object which stores 1024 points at a time in a buffer
    // which expands by 1024 more points whenever it gets filled up.  The data will
    // be written to a file named <aFileName> when the logger is deleted
    TheLogger = new CDataLogger (LOG_EXPANDING, 1024, aFileName);

    // Specify that the logger will save 5 columns of data, all doubles.  These
    // will hold the time and position and velocity of each conveyor axis
    TheLogger->DefineData (5, LOG_DOUBLE, LOG_DOUBLE, LOG_DOUBLE,
                              LOG_DOUBLE, LOG_DOUBLE);

    // Create data inbox to hold messages that say to turn logging off.  The inbox
    // will actually be owned by the parent task of this state
    CreateDataInbox ("Log Off");
    }
```

The destructor deletes the data logger object, freeing the memory it used and ensuring that the data is written to a disk file.

```
//-------------------------------------------------------------------------
// Destructor:  ~C_GlueDataLogger_Running
//     This destructor deletes the data logger object; this causes the logger to
//     automatically write its data to the data file.

C_GlueDataLogger_Running::~C_GlueDataLogger_Running (void)
    {
    delete TheLogger;
    }
```

The Action() function saves the data in real time. The process involves reading data supplied by the other tasks which generated that data, then asking the data logger to save the data by a call to the `SaveLoggerData()` function. The first parameter given to `SaveLoggerData()` is a pointer to the data logger object which is doing the saving. The other parameters specify the data being saved. These parameters *must* match both the number and type of data which was specified when the logger was configured by its `DefineData()` function. There is no way for the compiler to check whether the data fields match, so any errors in this parameter list will cause erroneous data to be saved.

```
//-------------------------------------------------------------------------
//  Function: Action
//      In this periodically executing function, the data logger stores data into its
//      data array in memory.

void C_GlueDataLogger_Running::Action (void)
    {
    double Position1, Position2;            // Positions of the conveyors
    double Velocity1, Velocity2;            //   and their velocities

    // Read the data inboxes which hold the data to be logged
    ReadGlobalData (BeltAxis1, "Position", &Position1);
    ReadGlobalData (BeltAxis1, "Velocity", &Velocity1);
    ReadGlobalData (BeltAxis2, "Position", &Position2);
    ReadGlobalData (BeltAxis2, "Velocity", &Velocity2);

    // Now save that data to the data logger object
    SaveLoggerData (TheLogger, (double)GetTimeNow (), Position1, Velocity1,
                                                     Position2, Velocity2);

    // This is a sample time task, so it must be idled or it will run all the time
    Idle ();
    }
```

The `TransitionTest()` function is used to cause a transition to another state when some condition is met. In the following function, the call to `ReadDataPack()` will return a `true` value when a signal has been sent by another task to cause the data logger to stop logging. In response to the receipt of a data package, the `TransitionTest()` function will return to the scheduler a pointer to the state to which a transition should occur. In this case, the pointer, `GDLStop`, points to the state object which represents a data logger which is not running. If no transition is to occur, the transition test function returns the value `NULL`, which represents a pointer which doesn't point to anything.

```
//-------------------------------------------------------------------------
//  Function: TransitionTest
//      This function checks to see if someone has told us to go to the stopped state.

CState* C_GlueDataLogger_Running::TransitionTest (void)
    {
    // If a message (any message) has appeared in the data inbox which signifies
    // time to stop logging, then remove that message and stop logging. Otherwise
    // just remain in this state
```

```
    if (ReadDataPack ("Log Off", NULL, NULL, NULL, NULL) == true)
        return GDLStop;
    else
        return NULL;
}
```

16.4.4 Data Communication Between Tasks

When a project has been partitioned into tasks and a set of task objects have been created, it is necessary for the tasks to communicate data among themselves. This data usually consists of variables whose values are to be transferred from one task to another. For example, if one task implements a PID controller and another implements a simulation of the process to be controlled (as is the case in module Glue02), the value of the actuation signal must be sent from the controller to the process, and the value of the process' output must be sent back to the controller.

In some cases, the regular methods of data exchange which are built into C++ can be utilized for intertask communication. These methods, which include calls to member functions and accessing statically allocated variables whose scope covers both of the tasks, can be used when both tasks are running in the same thread (i.e., synchronously with respect to one another) and in the same process. This is the case for most simulation-only programs and some simpler real-time programs. However, many real-time applications will take advantage of multithreading and multiprocessing environments, and the standard methods of data exchange will *not* work reliably under those circumstances.

In order to allow programs to work in multithreading and multiprocessing environments as well as single-thread and pure simulation modes, a more structured mechanism for transferring data between tasks must be used. The TranRun4 package implements a set of functions which belong to the task class `CTask` and are used to transfer data between tasks. These functions are also declared as member functions of the `CState` base class, so they can be used within a state object to transfer data to and from that state's parent task. The mechanisms by which data is transferred between tasks can vary depending on the environment in which the tasks are running; but these data transfer functions hide the process (network, shared memory, etc.) by which data is transferred so that the same user code can be run regardless of the environment.

There are two mechanisms for moving data across tasks, and they correspond roughly to the ideas of pushing and pulling data. The push method is used when a task has created some data which is needed by another task and wishes to send it. This is the data package mechanism. A data package implements a standardized format in which data is transferred. The data pack is defined as carrying the following items:

- An integer

- A double-precision floating point number

Section 16.4. Glue00: Conveyor Simulation

- A code specifying the task from which the data came

- A code for the process in which the sending task runs

This somewhat arbitrary format limits the types of data which can be communicated between tasks but results in a very simple protocol for sending the data and in a simple format for the functions which are used to send and receive the data. The actual meaning of the integer and floating point data items is left up to the user, so whatever data the user can encode into the available bits can be sent and received.

Data is transmitted by a call to the SendDataPack() function. This function will cause the data to be transmitted blindly toward the receiving task; that is, there is no way to tell if the data has actually been received. If confirmation is needed, one must have the receiving task send back a data package which confirms the receipt of the data. In addition to the integer and floating-point data being sent, SendDataPack() is given a parameter which determines if existing data in the receiving task's data package inbox is to be overwritten when the new data package arrives. The only two values which can be given are MB_OVERWRITE and MB_NO_OVERWRITE. The final parameter concerns a return receipt function which is currently not utilized. A typical call to SendDataPack() is shown here; it sends the integer 5 (which might be some sort of code) and some velocity data to the receiving task:

```
SendDataPack (TheOtherTask, "Speed", 5, Velocity, MB_OVERWRITE, MB_NO_RECEIPT);
```

If there is old data left in the receiving task's inbox, this function call will cause that old data to be overwritten. This is usually the right thing to do, because most control programs are concerned more with having the latest data available than with preserving old values.

A data package inbox is created by a call to the CreateDataInbox() function call. An inbox creation line from the belt axis constructor illustrates this procedure by example:

```
CreateDataInbox ("Voltage");
```

Because a receiving task may have many data inboxes, the name of the inbox is used to allow each data pack to be routed to the correct box. This means that the name used in the sending call to SendDataPack() must match the name used in creating the inbox and the name used to read the data from the inbox with the function ReadDataPack().[1] The data pack reading function reads data from an inbox and places it in local variables, if data is available. A typical call to this function is:

```
ReadDataPack ("Voltage", NULL, &Voltage, NULL, NULL);
```

---

[1] These names are case-sensitive, so Speed is *not* equal to speed; also, spaces and punctuation can be used in the names but these must match exactly as well.

This function call reads data from the inbox. The parameter &Voltage is a pointer to a local variable; if data has been placed into the data inbox since the inbox was last read, the new data will be placed in that local variable. If not, the contents of the variable will be unchanged. The parameters whose values are NULL are pointers to variables as well, but the NULL value indicates that the calling task isn't interested in those data items, so they will be discarded.

Data which is pulled by the receiving task is transmitted through the global shared database mechanism. The global database is simply a set of variables belonging to each task which are made available for other tasks to read. In the current implementation of TranRun4 only variables of type double may be used in the global database. As with data packages, the details of mutual exclusion and transmission across various media are handled internally within the scheduler. The use of the global database begins with a call to the function CreateGlobalData() in the constructor of the sending task:

```
CreateGlobalData ("Position", &Position);
```

Here the conveyor belt axis task is creating a data item which will make the belt's position available to other tasks. As with the data inboxes, the name of a global data item must match exactly that used in other tasks. The parameter &Position is the address of a variable which is to be communicated with other tasks.

The sending task will compute new values for the items in the global database within its Action() function and sometimes within the other state functions as well. After these new values have been computed, the function CopyGlobalData() is called. This function causes all the data in variables which have been associated with global data items through the CreateGlobalData() function to be copied into the global database. The copying process is protected from multithreading hazards by code within the scheduler. Now the data is available to be read by another task:

```
ReadGlobalData (BeltAxis1, "Position", &Position1);
```

In this function call, BeltAxis1 is the name of a conveyor belt axis object. When two or more belt axis objects have been created, global data from each one is read individually by using the pointer to each object. The ReadGlobalData() function always writes the most recently available data into the variable which is pointed to by its third argument; this is the data which was most recently written into the global database by a call to CopyGlobalData().

16.4.5 The Main File

In previous sections, the construction of task and state objects and the means by which they communicate with one another have been described. Now it's time to put these objects together into a working program. The main file is where all of the task and state objects are instantiated and configured and where the real-time scheduler is started.

Section 16.4. Glue00: Conveyor Simulation

Standard C and C++ programs have an entry point function called `main()`. This is the function which first runs when a program is invoked from the operating system. Modern GUI programs have similar functions, although their execution environment is more complex so that many entry and callback functions are needed. The TranRun4 scheduler provides a user entry point function called `UserMain()`. This function is called from within the scheduler's `main()` function immediately after the user's program is invoked. The first few lines of `UserMain()` are used to configure the scheduler and create a process object which will hold all the tasks:

```
int UserMain (int argc, char** argv)
{
// Set the master scheduler's timing parameters
TheMaster->SetTickTime (0.001);
TheMaster->SetStopTime (2.5);

// Create a process in which the tasks will run.  We use just one process for
// this example, running in one environment on one computer
CProcess* TheProcess = TheMaster->AddProcess ("The_Process");
...
```

A master scheduler object of class `CMaster` is automatically instantiated within the scheduler code. The master scheduler is responsible for starting and stopping the real-time program as well as measuring time and adjusting global parameters such as the tick time. The tick time serves different purposes depending on the real-time scheduling mode in which the scheduler is running. The Glue00 program is running in simulation mode, so the tick time is the amount of simulated time which the scheduler adds to the real-time clock value during each task scan. The stop time is the time at which the scheduler will cease running and return control to the `UserMain()` function.

A process object represents that part of the program which is running on one computer (or in one process on one computer for some environments). Simulation programs almost always run in one process; here that process is created by a call to the master scheduler's member function `AddProcess()` and a pointer to the process object is saved in `TheProcess`. This process pointer will be used when creating new tasks in the following section of the code from `UserMain()`:

```
// Instantiate a belt axis task, specifying a priority and sample time
BeltAxis1 = new CTask (TheProcess, "Belt Axis 1", SAMPLE_TIME, 10, 0.020);

// Create a state for the task.  Parameters: Task, Name, J, Kt, f, R, r
BeltAxis1_Running = new C_BeltAxis_Running (BeltAxis1, "Axis1run", 1.0, 2.5,
                                            0.1, 10.5, 0.1);
BeltAxis1->SetInitialState (BeltAxis1_Running);

// We have to send a message to the class telling it to set the current,
// so just ask the belt axis class to send the data package to itself
BeltAxis1->SendDataPack (BeltAxis1, "Voltage", 0, 3.0,
                    MB_OVERWRITE, MB_NO_RECEIPT);
```

```
// Repeat the belt axis creation process for another belt axis task
BeltAxis2 = new CTask (TheProcess, "Belt Axis 2", SAMPLE_TIME, 10, 0.020);
BeltAxis2_Running = new C_BeltAxis_Running (BeltAxis2, "Axis2run", 1.2, 2.3,
                                            0.15, 11.0, 0.1);
BeltAxis2->SetInitialState (BeltAxis2_Running);
BeltAxis2->SendDataPack (BeltAxis2, "Voltage", 0, 3.0,
                         MB_OVERWRITE, MB_NO_RECEIPT);
```

In this section of code we create two belt-axis tasks. First the task objects are constructed; as described in section 16.4.2, the task objects are instantiated from the base task class `CTask`. The call to the task object's constructors automatically cause the process to be informed of the existence of the new tasks. Then state objects are created. The states are objects of class `C_BeltAxis_Running`, a class which was derived from the standard state class `CState`. It is noteworthy that the two state objects `BeltAxis1_Running` and `BeltAxis2_Running` are created with different parameters to model conveyor belts whose physical properties are not quite the same. After the state objects have been created and automatically registered with the tasks which own them, the tasks are told which states are their initial states. The initial state is just the state in which a task will first run when real-time scheduling begins. Scheduling will begin with a call to the master scheduler's `Go()` method:

```
// Turn execution-time profiling on for all tasks and states
TheMaster->Go ();
```

After the real-time scheduler has finished running and returned control to the `UserMain()` function, some status information (discussed in section 16.5.1) is written and the program exits.

16.4.6 Glue00 Results

The Glue00 program runs the scheduler in simulated-time mode, and the data logger saves its data in a file called `GlueData.txt`. This file can be read into a spreadsheet or a mathematical analysis program such as Matlab. The latter was used to create the graphs of belt position and velocity shown in figure 16-2.

16.5 Glue01: An Oven Simulation

The glue curing oven is modeled with dynamic equations similar to those used for the conveyor belt model. The simplified equations which are used to model the oven are:

$$Q_{in} = i^2 R$$
$$Q_{out} = K(T_{oven} - T_{ambient})$$
$$\frac{dT}{dt} = \frac{Q_{in} - Q_{out}}{C_p}$$

Section 16.5. Glue01: An Oven Simulation

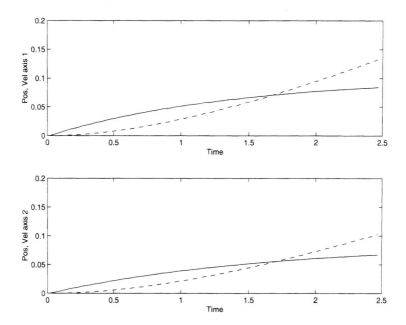

Figure 16-2. Belt axis simulation results

where Q_{in} is the heat transferred to the oven by current i flowing through resistance R, Q_{out} is the heat lost from the oven through thermal conductance K, C_p is the heat capacity of the oven, and dT/dt is the rate of change of the oven temperature.

These equations are implemented in the C_GlueOven_On class by the following code:

```
double HeatGenerated = Current * Current * CoilResistance;
double HeatLost = ThermCoeff * (OvenTemp - AmbientTemp);
OvenTemp += ((HeatGenerated - HeatLost) / HeatCapacity) * DeltaTime;
```

16.5.1 Configuration and Status Printouts

Some useful diagnostic functions are included with the TranRun4 package to make debugging real-time programs a little easier. The simplest of these are the configuration and status file writing functions. These functions belong to the CProcess class. The first writes a file which lists the tasks belonging to a process, along with the accompanying states, data inboxes, and global variables:

```
TheProcess->WriteConfiguration ("ProcessConfig.txt");
```

The configuration data file for the Glue01 task is shown in table 16.1.

Table 16.1. Configuration file printout

```
Configuration of Tasks in Process "The_Process"

Task                      Sample  State    Data      Global
Name            Type Pri. Time    Names    Inboxes   Data Items

[0]Belt Axis 1  Sample Time  10   0.020   Axis1run  Voltage   Velocity
                                                              Position

[1]Belt Axis 2  Sample Time  10   0.020   Axis2run  Voltage   Velocity
                                                              Position

[2]Glue Oven    Sample Time   6   0.100   Oven On             Temp
                                                              Current

[3]Data Logger  Sample Time   5   0.100   Running   Log Off
                                          Stopped   Log On
```

The second diagnostic printout function writes a file containing information about which tasks ran how many times and their status as the scheduler was stopped. This function is invoked as

`TheProcess->WriteStatus ("ProcessStatus.txt");`

and the resulting status data file for the Glue01 task is shown in table 16.2.

Table 16.2. Status file printout

```
Status Dump for Tasks in Process #1 "The_Process"
        Task: [0]Belt Axis 1      Type: Sample Time     Time: 96.846
        Status: Idle              Runs: 4843
        Sample Time: 0.02         Priority: 10

        Task: [1]Belt Axis 2      Type: Sample Time     Time: 96.846
        Status: Idle              Runs: 4843
        Sample Time: 0.02         Priority: 10

        Task: [2]Glue Oven        Type: Sample Time     Time: 96.846
        Status: Idle              Runs: 969
        Sample Time: 0.1          Priority: 6

        Task: [3]Data Logger      Type: Sample Time     Time: 96.846
        Status: Idle              Runs: 969
        Sample Time: 0.1          Priority: 5
```

16.6 Glue02: PID Control

The conveyor belt axes and the heater can both be controlled by standard single-input, single-output, PID controllers. The PID controllers are implemented as tasks. These controllers are designed to always run, so no state logic is needed for the control tasks. To use the PID control tasks, they must first be instantiated in the UserMain() file. The details of implementing the oven controller will be shown here; the belt axis PIDs are set up in the same way. Here is the oven controller's constructor call:

```
OvenPID = new C_SISO_PID_Controller (TheProcess, "Oven PID", 0.100, 6,
                         0.25, 0.05, 0.0);
```

Because the C_SISO_PID_Controller class is descended from the task class CTask, its constructor accepts the parameters which are given to a base task's constructor. These are the process in which the task runs, the name of the task, the sample time for the task, and the task's priority. The task type for a PID controller is internally set to type SAMPLE_TIME. In addition to the standard parameters, the C_SISO_PID_Controller constructor accepts three numbers which set its proportional, integral, and derivative control gains (K_p, K_i, and K_d).

Data flows into the controller by means of data inboxes. Inboxes are provided to allow parameter adjustment as well as for the exchange of setpoint and process data. The controller's inboxes are:

- "Setpoint"   The desired value of the process output
- "ProcVal"   The actual measured value of the process output
- "Kp"   Proportional control gain
- "Ki"   Integral control gain
- "Kd"   Derivative control gain
- "MaxActuation" Maximum value of actuation signal (for saturation)
- "MinActuation" Minimum value of actuation signal

There is just one global database item which is used to get information out of the controller:

- "Actuation"   The actuation signal which is sent to the controlled process

The oven is somewhat unusual as a control target in that its actuation signal is one-sided; that is, heat can be pumped into the heating coils, but heat cannot be pumped out (we assume that the oven lacks a refrigeration module). The PID controller's maximum and minimum actuation signal limits can be used to conveniently implement this sort of behavior. The code, taken from the UserMain() function in GlueMain.cpp, sets a maximum value for the heater current at 30 amps and a minimum at zero.

```
OvenPID->SendDataPack (OvenPID, "MaxActuation", 0, 30.0,
                       MB_OVERWRITE, MB_NO_RECEIPT);
OvenPID->SendDataPack (OvenPID, "MinActuation", 0, 0.0,
                       MB_OVERWRITE, MB_NO_RECEIPT);
```

In order to use the PID controller task, the oven task needs to be modified so that it keeps a pointer to the PID controller for use in sending and receiving data. The pointer is passed to the oven task in the constructor. This allows the creation of more ovens if needed, each with its own PID controller task.

```
//--------------------------------------------------------------------------------
// Constructor: C_GlueOven_On
//     The constructor saves parameters and parent task of this state

C_GlueOven_On::C_GlueOven_On (CTask* aTask, char* aName, double tAmb, double tCoef,
                              double coilR, double hCap, C_SISO_PID_Controller* aPID)
    : CState (aTask, aName)
    {
    AmbientTemp = tAmb;              // Save the parameters into member data
    ThermCoeff = tCoef;
    CoilResistance = coilR;
    HeatCapacity = hCap;
    ThePID = aPID;
```

Once the pointer **ThePID** has been sent to the oven task, it is used in the action function to send and receive data.

```
void C_GlueOven_On::Action (void)
    {
    // Get a new value for the simulated current
    ReadGlobalData (ThePID, "Actuation", &Current);

    ...control code is here...

    // Send the process value (temperature) back to the PID controller
    SendDataPack (OvenPID, "ProcVal", 0, OvenTemp, MB_OVERWRITE, MB_NO_RECEIPT);

    ...
    }
```

With these modifications, the oven task is run under PID control. The graphs in figure 16-3 show the oven's performance as it heats up to a temperature of 200 degrees. The conveyor belts are also equipped with PID controllers, and the resulting performance is shown in figure 16-4.

16.7 Glue03: The Operator Interface

This glue module adds a real-time operator interface to the project. Having an interface with which the user can interact as the program runs in real time is very helpful, not only for operating prototype and production systems but also for debugging. Watching the variables in the system change as the program runs,

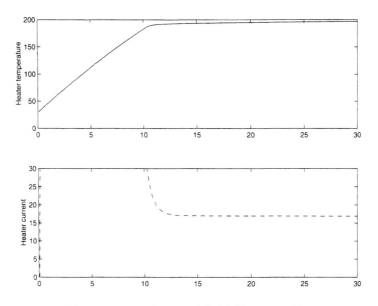

Figure 16-3. Oven with PID controller

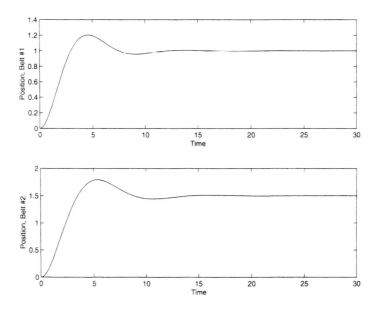

Figure 16-4. Conveyor belts with PID controller

and being able to change values in real time, can sometimes provide more valuable information about the operation of the system than diagnostic printouts.

The concept and implementation of the COperatorWindow class is discussed in chapter 9 and will not be repeated here. In this section the use of the operator interface within the glue project and details of using the interface within the TranRun4 environment will be discussed.

Most traditional interaction between computers and users is implemented through blocking code; for example, character input in standard C/C++ is performed by functions such as getchar(), which halt a program while waiting for the user to press a key. Character output can also cause large latencies; the time required to print a short string of text onto the character screen of a computer running MS-DOS, for example, can be a large fraction of a millisecond. This is unacceptable in a real-time environment.

The operator interface solves the latency problem by working within a task/state environment and using only nonblocking code. The operation of writing a screenful of data is broken down into many small bits, with each bit being written during one scan of a task's Action() function. Similarly, the keyboard is read by scanning for single keypresses on each pass through the action function and reading characters one at a time if they have already been typed.

This structure creates a one-to-one correspondence between functions in a TranRun4 program and functions in the operator interface class. The corresponding functions are:

Operator Interface Function	Location in TranRun4 Program
C_OperatorWindow constructor	Task/state constructor
Display()	Entry() function
Update()	Action() function
Close()	Exit code in TransitionTest() function

In addition, the operator interface is designed so that multiple windows showing different sets of information can be implemented in the project. These multiple windows correspond to states in the operator interface task.

The Glue03 module adds a two-window operator interface to the project. The first window, referred to as the main window because it appears when the program first begins running, allows control of the heater setpoint and displays the heater's current temperature. The second window allows control of the conveyor belt axes. The user can change the setpoints as well as the control gains K_p, K_I, and K_d for the two belts. The class definition for the main operator interface's state is shown here. The member data includes a pointer to the operator interface object as well as the variables to hold the data which is to be viewed and manipulated by the operator.[2]

---

[2] This member data is shared between two states. For a more detailed discussion of the storage of task data, see section 16.8.

Section 16.7. Glue03: The Operator Interface

```
class C_GlueOperInt_Main : public CState
    {
    private:
        COperatorWindow* TheOpWin;      // Pointer to operator interface object
        real_time TheTime;              // Time to be displayed on the screen
        CState* NextState;              // Next state to which to transition
        double OvenSetpoint;            // Temperature setpoint for oven
        double OvenTemperature;         // Current temperature from the oven

    public:
        C_GlueOperInt_Main (CTask*, char*);  // Standard state object constructor
        ~C_GlueOperInt_Main (void);          // Destructor to delete objects

        void Entry (void);              // Entry function initializes screen
        void Action (void);             // Action function updates it
        CState* TransitionTest (void);  // Test for transition to other state

        friend void ToBeltsWindow (void);  // This function causes a transition to
                                           // the other window; it's a "friend" so
    };                                     // it can access this state's member data
```

The operator interface task's constructor creates the interface object and assigns input, output, and key items. The constructor is:

```
C_GlueOperInt_Main::C_GlueOperInt_Main (CTask* pTask, char* aName)
    : CState (pTask, aName)
    {
    // Create the main operator window object
    TheOpWin = new COperatorWindow ("Glue System Main Window");

    // Add output items which display data in real time
    TheOpWin->AddOutputItem ("The Time", &TheTime);
    TheOpWin->AddOutputItem ("Oven Temp.", &OvenTemperature);

    // Add input items that allow the user to change data on the fly
    TheOpWin->AddInputItem ("Temp Setpt.", &OvenSetpoint);

    // Add key bindings so the user can press function keys and run functions
    TheOpWin->AddKey ('X', "eXit", "Program", StopEverything);
    TheOpWin->AddKey ('B', "Belt", "Window", ToBeltsWindow);

    OvenSetpoint = 200.0;               // Initialize the setpoint for the oven

    // Transitions to another state will only occur if asked for
    NextState = NULL;
    }
```

The **AddKey()** function associates a control key with a function. As long as this key function is defined at the beginning of the operator interface source file, the **AddKey()** function will recognize the name of the funtion as a function pointer. The function which is associated with the Ctrl-X key is shown here.

```
void StopEverything (void)
    {
    TheMaster->Stop ();
    }
```

Another key-mapped function is used to cause a transition from the main operator window state into the belt axis operator window state. Mapped to Ctrl-B, this function will change the `NextState` member variable of the main operator interface state object so that the next run of the `TransitionTest()` function causes a transition to the other window. This function must be declared a `friend` function of the `C_GlueOperInt_Main` class because the function must have access to the `NextState` variable, which is private class data.

```
void ToBeltsWindow (void)
    {
    OperInt_Main->NextState = OperInt_Belts;
    }
```

It is also important to remember to delete the operator interface object in the destructor of the operator interface task class:

```
C_GlueOperInt_Main::~C_GlueOperInt_Main (void)
    {
    delete TheOpWin;
    }
```

The `Entry()` and `Action()` functions house the operator interface's `Display()` and `Update()` functions respectively.

```
void C_GlueOperInt_Main::Entry (void)
    {
    TheOpWin->Display ();
    }

void C_GlueOperInt_Main::Action (void)
    {
    //  Get the current time so it can be proudly displayed
    TheTime = GetTimeNow ();

    //  Get data which is to be displayed from other tasks
    ReadGlobalData (GlueOven, "Temp", &OvenTemperature);

    //  Tell the operator window object to update the screen, check keys, etc.
    TheOpWin->Update ();

    //  Send the user-specified data to the other tasks which need it
    OvenPID->SendDataPack (OvenPID, "Setpoint", 0, OvenSetpoint,
                    MB_OVERWRITE, MB_NO_RECEIPT);
    }
```

A standard sequence of events is followed in the operator interface task's `Action()` function. First, the data which is to be displayed on the user screen is collected from the tasks which house it. In this case, the temperature is read from the oven task and the current time is found. Then the operator interface's `Update()` method is called. `Update()` may change the values of input items if the user has entered new data. Therefore, the last part of the action function sends the data which may have been changed by the user to the tasks which use that data. In this example, the user-controlled setpoint is sent to the oven controller task. It is possible to implement a smart update which only sends updated information to other tasks if the data within the user interface task has changed; however, data communication in most environments does not consume enough resources to make the extra work worthwhile.

16.7.1 Results

The main and belt axis operator interface windows display the text shown in table 16.3 and table 16.4.

The glue project operator interface permits the user to change the oven setpoint as the program is running. The data shown in figure 16-5 show the results of a simulation run during which the setpoint was changed from 200 to 120 degrees during the middle of the run. The results of actuator saturation combined with a somewhat high integral control gain can be seen.

When the belt axes are manipulated by the user in (simulated) real time, the results are as shown in figure 16-6. Here, the operator first moved the belt back and forth. Feeling somewhat dissatisfied with the performance of the controller, the operator then changed the control gains to an excessively high setting and again moved the belt back and forth. The results show a somewhat lackluster performance from the belt position controller; this issue will be addressed in the next section where an improved control scheme is implemented.

Table 16.3. Main glue operator window

```
Glue System Main Window

Temp Setpt.    ->              200         The Time     114.86
                                           Oven Temp.   199.958




















Ctrl-X    Ctrl-B
eXit      Belt
Program   Window
```

Table 16.4. Conveyor belt axis control window

```
Glue System Belt Axis Control Window

Belt 1 Setpt              4            The Time     145.86
Belt 1 Kp                 20           Belt 1 Pos   4.003
Belt 1 Ki       ->        0.02         Belt 2 Pos   0
Belt 1 Kd                 0
Belt 2 Setpt              0
Belt 2 Kp                 10
Belt 2 Ki                 0.03
Belt 2 Kd                 0




Ctrl-X      Ctrl-W
eXit        Main
Program     Window
```

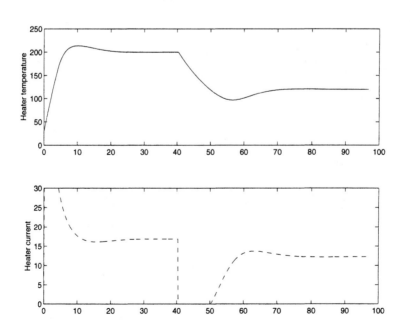

Figure 16-5. Oven with PID and changing setpoint

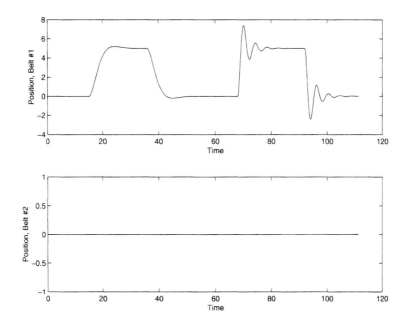

Figure 16-6. Changing setpoints and gains for belt

16.8 Glue04: Motion Profiling

In order to improve the less than optimal performance of the position controller shown in Glue modules 02 and 03, a motion profiler is implemented in this section. The profiler will cause the belt axes to move in a predefined pattern of smooth acceleration and deceleration, rather than just handing a step input to a PID controller and letting the parts fall where they may (and fall they may, right off the belt, if the acceleration is too great).

The operation of the profiler is theoretically simple. The conveyor belt is accelerated from the starting point toward a given target point at a constant rate until it reaches a predetermined maximum speed. The acceleration and maximum speed are determined by equipment constraints or safety concerns. The belt then cruises at the maximum speed until it has reached a point at which a steady deceleration from the cruise speed to zero speed will result in the belt reaching the target point just as it decelerates to zero speed. Once stopped at the target point, the belt is placed in a hold mode in which a PID controller is used to maintain its position at the target position in the presence of disturbances.

To this simple scheme are added a few complications on the way to implementation. If the starting point and target point are sufficiently close together, there will not be time to reach the cruising velocity before deceleration to zero speed must begin. The finite and slightly variable sample time of the profiling and control tasks also has an effect on the profiler's operation. These effects can be mitigated by a careful choice of sampling times for the simulation, control, and motion profiling tasks.

The movement of the belt is controlled by moving the position setpoint of the same PID controller which has been used in previous Glue modules. The controller then tracks the motion of the setpoint, compensating for disturbances as it does so. When the setpoint's position x is accelerating at a constant rate, its velocity v is increasing linearly: $v = at$. The position is then given by $x = vt$ or $x = at^2/2$. These equations are written in difference form to be placed into the program:

```
Velocity += CurrentAccel * GetSampleTime ();
ClipToBounds (Velocity, CruiseSpeed);
Xcurrent += Velocity * GetSampleTime ();
```

Here, `Xcurrent` is the current position of the setpoint. The function `GetSampleTime()` returns the nominal time between executions of the task. This use assumes that the motion profiling task really is being executed at a rate equal to its sample time (on average). If it is suspected that this sample time might often not be met, it would be better to measure the sample time for use in these calculations. An equation which clips the velocity by limiting it to a maximum value given in `CruiseSpeed` is implemented by the macro `ClipToBounds()`. If the velocity is set to a value higher than the cruise velocity, `ClipToBounds()` resets the velocity to the cruise velocity limit.

Ideally, the system should begin deceleration when the belt is at a distance $x = -at^2/2$ from the target point. However, it is unlikely that the motion task will be executing precisely at the time when deceleration must begin, and errors in summation (as opposed to true integration) will most likely cause the actual deceleration to differ from the ideal. The result is that the setpoint will stop at a position some distance away from the target point. This effect can be minimized by constantly recalculating the deceleration as the motion is slowed down. This repeated recalculation of the corrected acceleration a_c follows the formula $a_c = -v|v|/(2d)$. The use of the absolute value of v causes the acceleration to be positive or negative depending on the direction of the distance in which the belt is travelling. The code which implements this scheme is shown below.[3]

```
Adjust = -(Velocity * Velocity) / (2.0 * (Xtarget - Xcurrent));
Velocity += Adjust * GetSampleTime ();
ClipToBounds (Velocity, CruiseSpeed);
Xcurrent += Velocity * GetSampleTime ();
```

The task code in which these equations are placed has four states; the state transition diagram is shown in figure 16-7. The normal sequence of actions would occur as follows: The profiler begins in either the `Idle` or `Hold` state with the conveyor not moving. A command arrives to move to a new target point. The

---

[3] The variables used in this function are member data of the *task* class C_MotionP and are accessed by using a pointer to that class. These pointers have been removed for clarity.

Section 16.8. Glue04: Motion Profiling

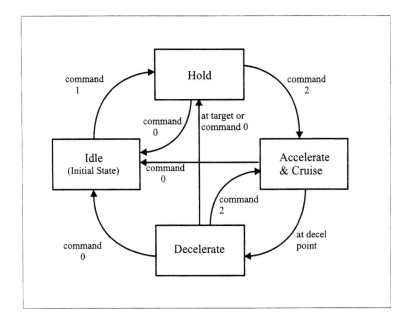

Figure 16-7. Transition logic for motion profiler task

profiler enters the Accelerate state and the belt begins to move. At some point, the point is reached at which deceleration should begin, and a transition to the Decelerate state occurs. When the belt has come sufficiently close to the target point and slowed to a near stop, the profiler transitions to the Hold state in which the PID controller is given a fixed setpoint at which the belt will stay until a new command to move is given.

The profiler receives commands in a data inbox named Command. The floating-point data item holds a target point to which the profiler should move, and the integer data item holds a command. Valid commands are 0 to turn control off, 1 to cause the profiler to hold the belt fixed at the given point, and 2 to perform a profiled move. The status of the profiler is available in a global data item called Status; this floating point number is given the value 0.0 when the profiler is in the Idle state, 1.0 for the Hold state, 2.0 for acceleration and 3.0 for deceleration. Another task which needs to synchronize its activities with those of the profiler can use these values to determine what the profiler is up to. It is assumed that the PID controller and belt hardware perform sufficiently well that large deviations from the setpoint do not occur.

The state transition logic structure in figure 16-7 is the first such structure in the Glue project which has many states with multiple transitions. This structure is easily implemented using the same techniques which have been used in previous modules, but the TransitionTest() functions are more complex. The body of the test function which causes exit from the Hold state is shown here.

```
CMotionP_Hold::TransitionTest( )
    int Command = 0;                        // Code sent as part of a data pack

    // If a command data pack was received, figure out what the contents mean
    if (ReadDataPack ("Command", &Command, &Xtarget, NULL, NULL) == true)
    {
        if (Command == 0) return (IdleState);
        else if (Command == 2) return (AccelState);
        // Any other command is 1 (remain in this state) or invalid and is ignored
    }
    // There was no valid command to change states, so stay in this one
    return (NULL);
```

The only exit from the Hold state occurs when a command arrives to enter the Idle state or to enter the Accelerate state and begin moving toward a new target point. The Accelerate state can be exited if a command is received, but it must also be exited when the time to decelerate has arrived. The body of the transition test function which performs these tests is shown next; the other transition test functions work in a similar manner.

```
CMotionP_Accel::TransitionTest( )
// First check if a message has come in ordering the controller be idled or
// a new target point has been selected
if (ReadDataPack ("Command", &Command, &Xtarget, NULL, NULL) == true)
    {
    // A command integer of 0 means to idle profiler and turn off controller
    if (Command == 0) return (IdleState);

    // A command of 1 means to go and hold the given target point
    if (Command == 1) return (HoldState);

    // A 2 command means to move there on a new profile, so re-enter this state
    if (Command == 2) return (AccelState);

    // Any other command is invalid and just gets ignored, so there
    }

// No command has come in.  Check where the current acceleration is going to
// put us; if we're near overshooting the target point, begin deceleration
double Xstop = StopPoint (Xcurrent, Velocity, CurrentAccel);
if ((Direction > 0.0) && (Xstop >= Xtarget)) return (DecelState);
if ((Direction < 0.0) && (Xstop <= Xtarget)) return (DecelState);

return (NULL);
}
```

Storage of Task Data

In previous one-state tasks, data has been stored in member variables of the single state. This is a good place to keep data because state member data is easily accessible within the state's member functions, because the data is kept private within the state to prevent accidental manipulation, and because a new copy of state member data is created for each state object which is created. Thus, instantiating two belt axis simulation objects creates two independent tasks, each with its own protected data.

The storage of data becomes less convenient when more states are added. The data must now be made accessible to all the states belonging to a class. Several methods are available to do this, each with its own advantages and disadvantages.

- The data can be declared as static variables at the top of the task's source (.cpp) file. This makes access to the data easy for all the task and state member functions and restricts access to only the functions within the task's source file. However, in general one *cannot* create more than one task of this class because only one copy of the static member data will be created and shared between instantiations of the task class. This virtually guarantees unpredictable side effects.

- The data can be declared as member data of one of the states, and the others can use a pointer to the first state to access the data. The other states are declared as friend classes to the state holding the data. If more than one instance of the task is to be created, however, care must be taken to insure that the pointer to the state which owns the data which is given to each other state object. This is not a global pointer but a pointer which belongs to the task class and is thus distinct for each instance of the task. This method can also create somewhat strange looking code in which one state accesses task data without pointers while the others use pointers.

- A custom task class can be created as a container for the task data, and all the states are declared as friend classes of the task class. States then use a built-in pointer to their parent state, pParent, to access data items belonging to the task class. Since each task object is instantiated independently of other tasks, this method ensures separate copies of the data for every task. It requires the extra effort of creating a new task class, and most data items accessed within the states must use the clumsy notion of pParent->Data.

The method of using the task class to store task data is chosen for the motion profiler task. The task class declaration is shown here.

```
class C_MotionP : public CTask
    {
    private:
        C_SISO_PID_Controller* ThePID;      // PID controller used by this profiler
        double Xcurrent;                    // Current position of the setpoint
        double Xtarget;                     // Position to which we're moving
        ...more member data...
        double Velocity;                    // Current velocity of setpoint motion
        double Direction;                   // Direction of travel, 1 or -1
```

```
        C_MotionP_Idle* IdleState;         // Here we declare pointers to all the
        C_MotionP_Hold* HoldState;         // state objects which belong to this
        C_MotionP_Accel* AccelState;       // task class
        C_MotionP_Decel* DecelState;

        double StopPoint (double, double,  // Function to find where a
                         double);          // hypothetical profile would stop

    public:
        // Constructor just calls the sample time task's constructor
        C_MotionP (CProcess*, char*, int, real_time, C_SISO_PID_Controller*,
                   double, double);

        // Make all the states friend classes so they can get at this task's data
        friend class C_MotionP_Idle;
        friend class C_MotionP_Hold;
        friend class C_MotionP_Accel;
        friend class C_MotionP_Decel;
};
```

Because pointers to the state classes must be kept as task member data, it is convenient to instantiate the state objects within the task class's constructor. This method has the advantage of requiring fewer constructor calls in UserMain(). The constructor code which creates the state objects is as follows.

```
C_MotionP::C_MotionP (CProcess* aProc, char*aName, int aPri, real_time aTime,
                     C_SISO_PID_Controller* aPID, double aSpeed, double aAccel)
    : CTask (aProc, aName, SAMPLE_TIME, aPri, aTime)
    {
    // Save the parameter
    ThePID = aPID;

    // Create the member states of this task and specify one as the initial state
    IdleState = new C_MotionP_Idle (this, "Idle");
    HoldState = new C_MotionP_Hold (this, "Holding");
    AccelState = new C_MotionP_Accel (this, "Accel");
    DecelState = new C_MotionP_Decel (this, "Decel");
    SetInitialState (IdleState);

    //...more code...

    CreateDataInbox ("Command");

    //...more boxes...

    CreateGlobalData ("Velocity", &Velocity);
    }
```

Profiled Motion Results

The completed motion profiler produces smooth motion which is much less likely to drop parts on the floor than that from the previous Glue module. A profiled motion graph is shown in Figure 16-8. The belt was commanded by a slightly modified user interface to move from a position of 0 to 5, and after it had stopped and entered the Hold state, a move from 5 to 1 was ordered. Slight overshoot due to the integral action in the PID controller can be seen. The positions shown are taken from the simulation; the velocity is that of the setpoint rather than that of the simulated conveyor belt.

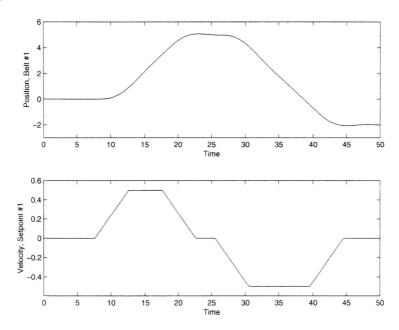

Figure 16-8. Transition logic for motion profiler task

The Transition Logic Trace

The process of testing and debugging a program often takes more time than all other aspects of a project combined. Useful debugging tools can save a great deal of that time. One tool which is provided with the TranRun4 package is the transition logic trace.

The transition logic trace is a record of when each of the tasks in your project executed a transition from one state to another. It is also referred to as a transition audit trail. The trace is automatically recorded by the scheduler each time a program is run. After the real-time portion of the program has finished running, a call to the WriteTrace() method of the master scheduler will cause the trace to be written to a disk file as follows:

```
TheMaster->WriteTrace ("TL_Trace.txt");
```

The following transition logic trace was made during a brief run of the Glue04 program. The operator switched the user interface screen from the main screen to the belt control screen and then caused one belt axis to execute a profiled move to a new target point. After the belt axis had finished its motion, the operator switched from the belt control interface screen to the main screen and exited the program.

```
TranRun4 State Transition Logic Trace File

Time       Process        Task           From            To

2.3853     The_Process    Glue OpInt     Main OpInt      Belts OpInt
10.6829    The_Process    Axis 1 Prof    Idle            Accel
16.6834    The_Process    Axis 1 Prof    Accel           Decel
21.8842    The_Process    Axis 1 Prof    Decel           Holding
29.0548    The_Process    Glue OpInt     Belts OpInt     Main OpInt
```

The information in the transition logic trace can often be very helpful in debugging. Because the design of a program is embodied in its transition logic structure, the most important information about whether the program did what it was designed to do (and when it did so) is contained in the transition logic trace.

16.9 Glue05: Belt Sequencing

In the first 5 modules, several tasks have been created to manage the behavior of the individual belt axes. However, the final control system will require multiple layers of supervison and coordination between these lower-level tasks. This module introduces the `Transport` task, which will eventually act as a midlevel supervisor of both the belt motion and the unloading robot. At this point, the `Transport` task is only responsible for managing tasks which represent clamps that secure the assembly to the belt. Its executable code is accordingly simple, with each state serving basically to send a command signal to the appropriate clamp task.

The `Clamp` tasks themselves are minimal simulations of clamp operation; no dynamic model is included in the task code, but rather fixed time intervals represent the opening and closing of the clamp. A transition logic trace file for this module is:

```
TranRun4 State Transition Logic Trace File

Time       Process        Task           From            To

0.0200     Process_One    Transport 2    Start           Close
0.0400     Process_One    Transport 1    Start           Close
0.0700     Process_One    Clamp 2        Clamp Open      Clamp Closing
0.0900     Process_One    Clamp 1        Clamp Open      Clamp Closing
0.6300     Process_One    Clamp 2        Clamp Closing   Clamp Closed
0.6500     Process_One    Clamp 1        Clamp Closing   Clamp Closed
0.7000     Process_One    Transport 2    Close           Wait 1
0.7200     Process_One    Transport 1    Close           Wait 1
3.2600     Process_One    Transport 2    Wait 1          Open
3.2800     Process_One    Transport 1    Wait 1          Open
3.3100     Process_One    Clamp 2        Clamp Closed    Clamp Opening
3.3300     Process_One    Clamp 1        Clamp Closed    Clamp Opening
3.6700     Process_One    Clamp 2        Clamp Opening   Clamp Open
```

3.6900	Process_One	Clamp 1	Clamp Opening	Clamp Open
3.7400	Process_One	Transport 2	Open	Wait 2
3.7600	Process_One	Transport 1	Open	Wait 2
5.3800	Process_One	Transport 2	Wait 2	Stop
5.4000	Process_One	Transport 1	Wait 2	Stop

For this module, each `Transport` task is implemented as a separate class, to illustrate a situation where the code for each might be very different. However, this task will evolve over the next two examples to become more complex, while also being reduced to a single class definition for generality.

One note on syntax should be made at this point. In the `C_Clamp` class, several static values are defined for use in global data and message passing. For example, the static member `C_Clamp::OPEN_CLAMP` is intended to be a value accepted by the Clamp Command message inbox of a clamp task instance. By defining such integer commands as part of the class, and referring to only these variables when doing intertask messaging, the effect of the messages becomes independent of the actual values. In essence, it doesn't matter whether `OPEN_CLAMP` is set to 1 or 2001: the result when sending the value to the message inbox will be the right one. Any other task can use static values through the scope resolution operator and the class name.

16.10 Glue06: The Glue Application Machine

The overall gluing simulation involves more modules than just the belt axes and oven. The complete system includes robots to load parts, apply glue, and unload the finished assemblies. However, including dynamic models of all these components at this stage overly complicates matters. The complete control program can be developed using preset time intervals to represent robot motion, just as in the previous Glue example. This allows program flow to be designed and debugged without spending a lot of time dealing with numerics. Once the interactions of the tasks have all been established, appropriate numerical simulation code can be added at a later stage without changing the high-level interface.

In addition to the three minimal robot simulation tasks, the main supervisory task is also introduced in this Glue module. The `Assembly` task will eventually coordinate operation of the loading robot, the glue applicator robot, and each of the `Transport` tasks. Initially, however, the `Assembly` task will only simluate a single loading/gluing cycle. The `Transport` task has been modified to open and close the associated clamps in the proper order for the final process, and has also integrated an associated unloading robot.

All of the coordination between tasks is done by sending and receiving task messages, with very little reliance on global data values. This is necessary in order to ensure that critical commands are not missed. For the most part, signals are sent from the supervisory tasks to those at a lower level, which in turn respond once the requested operation is complete. In order for the lower-level tasks (for example, those simulating the loading or unloading robots) to remain isolated from those controlling them, the task classes include methods to register the task and message box where a response should be sent.

The following is a transition logic trace file for this module:

```
TranRun4 State Transition Logic Trace File

Time        Process         Task              From                  To

0.0100      Process_One     Assembly          Start                 Place Base
0.0200      Process_One     Transport 2       Start                 Close
0.0300      Process_One     Unload Robot 2    Start                 Ready
0.0500      Process_One     Transport 1       Start                 Close
0.0600      Process_One     Unload Robot 1    Start                 Ready
0.0800      Process_One     Glue Applicator   Start                 Ready
0.0900      Process_One     Load Robot        Start                 Ready
0.1300      Process_One     Clamp 2           Clamp Open            Clamp Closing
0.1600      Process_One     Clamp 1           Clamp Open            Clamp Closing
0.1800      Process_One     Load Robot        Ready                 Base Pickup
0.7600      Process_One     Clamp 2           Clamp Closing         Clamp Closed
0.7900      Process_One     Clamp 1           Clamp Closing         Clamp Closed
0.9200      Process_One     Transport 2       Close                 Wait 1
0.9500      Process_One     Transport 1       Close                 Wait 1
2.0700      Process_One     Load Robot        Base Pickup           Base Placement
2.4500      Process_One     Transport 2       Wait 1                Open
2.5600      Process_One     Clamp 2           Clamp Closed          Clamp Opening
3.0100      Process_One     Clamp 2           Clamp Opening         Clamp Open
3.1700      Process_One     Transport 2       Open                  Wait 2
3.5600      Process_One     Transport 1       Wait 1                Open
3.6700      Process_One     Clamp 1           Clamp Closed          Clamp Opening
3.9600      Process_One     Load Robot        Base Placement        Moving To Ready
4.1200      Process_One     Clamp 1           Clamp Opening         Clamp Open
4.2800      Process_One     Transport 1       Open                  Wait 2
5.9900      Process_One     Transport 1       Wait 2                Unload Assembly
6.0800      Process_One     Transport 1       Unload Assembly       Wait Until Clear
6.0900      Process_One     Unload Robot 1    Ready                 Assembly Pickup
6.5700      Process_One     Load Robot        Moving To Ready       Ready
6.6700      Process_One     Assembly          Place Base            Apply Glue
6.8300      Process_One     Glue Applicator   Ready                 Moving To Glue
7.3100      Process_One     Transport 2       Wait 2                Unload Assembly
7.4000      Process_One     Transport 2       Unload Assembly       Wait Until Clear
7.4100      Process_One     Unload Robot 2    Ready                 Assembly Pickup
8.2500      Process_One     Unload Robot 1    Assembly Pickup       Assembly Placement
8.4200      Process_One     Transport 1       Wait Until Clear      Wait 3
9.5700      Process_One     Unload Robot 2    Assembly Pickup       Assembly Placement
9.7400      Process_One     Transport 2       Wait Until Clear      Wait 3
10.4100     Process_One     Unload Robot 1    Assembly Placement    Moving To Ready
10.4300     Process_One     Glue Applicator   Moving To Glue        Gluing
11.7300     Process_One     Unload Robot 2    Assembly Placement    Moving To Ready
12.7500     Process_One     Unload Robot 1    Moving To Ready       Ready
14.0700     Process_One     Unload Robot 2    Moving To Ready       Ready
20.9600     Process_One     Glue Applicator   Gluing                Moving To Ready
24.5600     Process_One     Glue Applicator   Moving To Ready       Ready
24.6700     Process_One     Assembly          Apply Glue            Place Cylinder
24.8400     Process_One     Load Robot        Ready                 Cylinder Pickup
26.5500     Process_One     Load Robot        Cylinder Pickup       Cylinder Placement
28.2600     Process_One     Load Robot        Cylinder Placement    Moving To Ready
30.8700     Process_One     Load Robot        Moving To Ready       Ready
30.9700     Process_One     Assembly          Place Cylinder        Wait
33.5300     Process_One     Transport 1       Wait 3                Stop
```

Note that the `Assembly` and `Transport` tasks are pursuing completely independent behavior at this point. However, careful reading of this trace file will reveal that the proper sequence of operations is taking place. The `Assembly` task first instructs the loading robot to load the base, after which the glue applicator robot applies the glue. After the load robot places the cylinder, the high-level assembly task is finished. Each `Transport` task first opens and closes its clamps, followed by the movement of the appropriate unloading robot. Of course, the operations of these two supervisory layers will be fully coordinated by the final example.

16.11 Glue07: Transport Task Supervision

Although `Clamp` task and unloading robot control have already been integrated into the `Transport` task, it still does not send any belt movement commands. In the final revision of the `Transport` task, supervision of one motion profiler task is added and the state structure changed slightly to reflect the exact transport sequence. Of course, the `Transport` task will now need to know when the profiler (and hence, the belt) has completed its assigned motion. To provide this response, the `SetNotifyTask()` function is added to the motion profiler task class. The simulation of the transport supervisor will then reflect the dynamic response of the belt axis, along with the time interval simulation of the unloading robot.

The final state transition sequence of the transport task encompasses one entire movement of the assembly. First, the clamps are moved to the loading location. After gluing, the clamps are closed and the assembly is passed through the oven. In this simulation, the curing process is represented by a fixed time interval. Finally, the assembly is moved to the unloading location, where the associated unload robot is signalled to come pick it up. As soon as the assembly is removed, the transport task moves the belt back to its starting position, allowing the assembly to be moved by the unloading robot at the same time.

A transition logic trace file for this example is shown next. Note that all of the loading portions of the assembly task have been turned off for this example, and that only the operation of each transport task is being tested. Figure 16-9 shows the position of each belt axis during the course of the test.

```
TranRun4 State Transition Logic Trace File

Time        Process         Task                From                To

    0.0110  Process_One     Assembly            Start               Test One
    0.0220  Process_One     Transport 2         Start               Moving To Ready
    0.0330  Process_One     Unload Robot 2      Start               Ready
    0.0550  Process_One     Transport 1         Start               Moving To Ready
    0.0660  Process_One     Unload Robot 1      Start               Ready
    0.0880  Process_One     Glue Applicator     Start               Ready
    0.0990  Process_One     Load Robot          Start               Ready
    8.5360  Process_One     Transport 2         Moving To Ready     Ready
    8.5690  Process_One     Transport 1         Moving To Ready     Ready
    8.6680  Process_One     Transport 1         Ready               Open 1
```

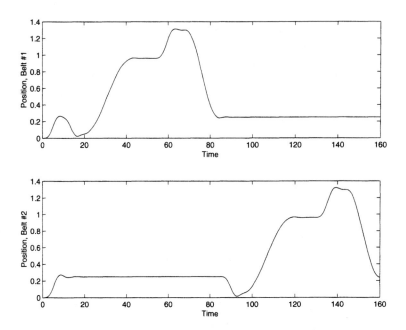

Figure 16-9. Conveyor belt positions

8.7670	Process_One	Transport 1	Open 1	Moving To Load
16.5880	Process_One	Transport 1	Moving To Load	Close
16.7090	Process_One	Clamp 1	Clamp Open	Clamp Closing
17.4020	Process_One	Clamp 1	Clamp Closing	Clamp Closed
17.5780	Process_One	Transport 1	Close	Moving To Oven
43.3180	Process_One	Transport 1	Moving To Oven	Curing
53.5150	Process_One	Transport 1	Curing	Moving To Unload
63.4150	Process_One	Transport 1	Moving To Unload	Open 2
63.5360	Process_One	Clamp 1	Clamp Closed	Clamp Opening
64.0310	Process_One	Clamp 1	Clamp Opening	Clamp Open
64.2070	Process_One	Transport 1	Open 2	Unload Assembly
64.3170	Process_One	Unload Robot 1	Ready	Assembly Pickup
66.4950	Process_One	Unload Robot 1	Assembly Pickup	Assembly Placement
66.6820	Process_One	Transport 1	Unload Assembly	Moving To Ready
68.6730	Process_One	Unload Robot 1	Assembly Placement	Moving To Ready
71.0490	Process_One	Unload Robot 1	Moving To Ready	Ready
84.2050	Process_One	Transport 1	Moving To Ready	Ready
84.3590	Process_One	Assembly	Test One	Test Two
84.4690	Process_One	Transport 2	Ready	Open 1
84.5680	Process_One	Transport 2	Open 1	Moving To Load
92.3890	Process_One	Transport 2	Moving To Load	Close
92.5100	Process_One	Clamp 2	Clamp Open	Clamp Closing
93.2030	Process_One	Clamp 2	Clamp Closing	Clamp Closed
93.3790	Process_One	Transport 2	Close	Moving To Oven
119.1190	Process_One	Transport 2	Moving To Oven	Curing

129.3160	Process_One	Transport 2	Curing	Moving To Unload
139.2160	Process_One	Transport 2	Moving To Unload	Open 2
139.3370	Process_One	Clamp 2	Clamp Closed	Clamp Opening
139.8320	Process_One	Clamp 2	Clamp Opening	Clamp Open
140.0080	Process_One	Transport 2	Open 2	Unload Assembly
140.1180	Process_One	Unload Robot 2	Ready	Assembly Pickup
142.2960	Process_One	Unload Robot 2	Assembly Pickup	Assembly Placement
142.4830	Process_One	Transport 2	Unload Assembly	Moving To Ready
144.4740	Process_One	Unload Robot 2	Assembly Placement	Moving To Ready
146.8500	Process_One	Unload Robot 2	Moving To Ready	Ready
160.0060	Process_One	Transport 2	Moving To Ready	Ready
160.1930	Process_One	Assembly	Test Two	Stop

16.12 Glue08: The Completed Assembly System

With the Transport task completed, all that remains is to construct the proper sequence for the Assembly task. Of course, the number of assemblies to complete must also be specified at the beginning of program execution. The addition of states appropriate to sequence startup and shutdown, as well as one to test/wait for an available transport task, quickly completes the final Assembly task. A transition logic trace file for the completed simulation is shown, as well as a plot of the belt axis positions in figure 16-10.

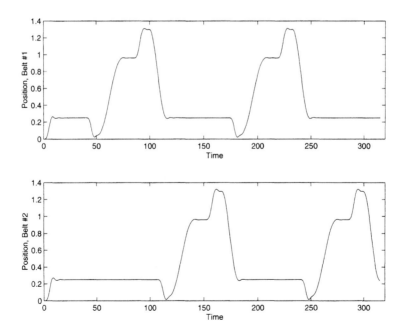

Figure 16-10. Conveyor belt positions

TranRun4 State Transition Logic Trace File

Time	Process	Task	From	To
0.0220	Process_One	Transport 2	Start	Moving To Ready
0.0550	Process_One	Transport 1	Start	Moving To Ready
8.5360	Process_One	Transport 2	Moving To Ready	Ready
8.5690	Process_One	Transport 1	Moving To Ready	Ready
8.7230	Process_One	Assembly	Start	Begin New
8.8220	Process_One	Assembly	Begin New	Place Base
15.5540	Process_One	Assembly	Place Base	Apply Glue
33.7700	Process_One	Assembly	Apply Glue	Place Cylinder
40.1060	Process_One	Assembly	Place Cylinder	Wait for Transport
40.2050	Process_One	Assembly	Wait for Transport	Cure/Unload
40.3480	Process_One	Transport 1	Ready	Open 1
40.4470	Process_One	Transport 1	Open 1	Moving To Load
48.2680	Process_One	Transport 1	Moving To Load	Close
49.2580	Process_One	Transport 1	Close	Moving To Oven
74.9980	Process_One	Transport 1	Moving To Oven	Curing
75.1520	Process_One	Assembly	Cure/Unload	Begin New
75.2510	Process_One	Assembly	Begin New	Place Base
81.9830	Process_One	Assembly	Place Base	Apply Glue
85.1950	Process_One	Transport 1	Curing	Moving To Unload
95.0950	Process_One	Transport 1	Moving To Unload	Open 2
95.8870	Process_One	Transport 1	Open 2	Unload Assembly
98.3620	Process_One	Transport 1	Unload Assembly	Moving To Ready
100.1990	Process_One	Assembly	Apply Glue	Place Cylinder
106.5350	Process_One	Assembly	Place Cylinder	Wait for Transport
106.6340	Process_One	Assembly	Wait for Transport	Cure/Unload
106.7440	Process_One	Transport 2	Ready	Open 1
106.8430	Process_One	Transport 2	Open 1	Moving To Load
114.6640	Process_One	Transport 2	Moving To Load	Close
115.6540	Process_One	Transport 2	Close	Moving To Oven
115.8850	Process_One	Transport 1	Moving To Ready	Ready
141.3940	Process_One	Transport 2	Moving To Oven	Curing
141.5810	Process_One	Assembly	Cure/Unload	Begin New
141.6800	Process_One	Assembly	Begin New	Place Base
148.4120	Process_One	Assembly	Place Base	Apply Glue
151.5910	Process_One	Transport 2	Curing	Moving To Unload
161.4910	Process_One	Transport 2	Moving To Unload	Open 2
162.2830	Process_One	Transport 2	Open 2	Unload Assembly
164.7580	Process_One	Transport 2	Unload Assembly	Moving To Ready
166.6280	Process_One	Assembly	Apply Glue	Place Cylinder
172.9640	Process_One	Assembly	Place Cylinder	Wait for Transport
173.0630	Process_One	Assembly	Wait for Transport	Cure/Unload
173.2060	Process_One	Transport 1	Ready	Open 1
173.3050	Process_One	Transport 1	Open 1	Moving To Load
181.1260	Process_One	Transport 1	Moving To Load	Close
182.1160	Process_One	Transport 1	Close	Moving To Oven
182.2810	Process_One	Transport 2	Moving To Ready	Ready
207.8560	Process_One	Transport 1	Moving To Oven	Curing
208.0100	Process_One	Assembly	Cure/Unload	Begin New
208.1090	Process_One	Assembly	Begin New	Place Base
214.8410	Process_One	Assembly	Place Base	Apply Glue
218.0530	Process_One	Transport 1	Curing	Moving To Unload

227.9530	Process_One	Transport 1	Moving To Unload	Open 2
228.7450	Process_One	Transport 1	Open 2	Unload Assembly
231.2200	Process_One	Transport 1	Unload Assembly	Moving To Ready
233.0570	Process_One	Assembly	Apply Glue	Place Cylinder
239.3930	Process_One	Assembly	Place Cylinder	Wait for Transport
239.4920	Process_One	Assembly	Wait for Transport	Cure/Unload
239.6020	Process_One	Transport 2	Ready	Open 1
239.7010	Process_One	Transport 2	Open 1	Moving To Load
247.5220	Process_One	Transport 2	Moving To Load	Close
248.5120	Process_One	Transport 2	Close	Moving To Oven
248.7430	Process_One	Transport 1	Moving To Ready	Ready
274.2520	Process_One	Transport 2	Moving To Oven	Curing
274.4390	Process_One	Assembly	Cure/Unload	Begin New
274.5380	Process_One	Assembly	Begin New	Stop
284.4490	Process_One	Transport 2	Curing	Moving To Unload
294.3490	Process_One	Transport 2	Moving To Unload	Open 2
295.1410	Process_One	Transport 2	Open 2	Unload Assembly
297.6160	Process_One	Transport 2	Unload Assembly	Moving To Ready
315.1390	Process_One	Transport 2	Moving To Ready	Ready

There is one notable feature of this example, involving the overlapped operation of the different assembly operations. Evidence of this can be seen starting at time = 75.1520, when the assembly task begins to coordinate the gluing of a second widget while the first is still in the oven. Subsequently, at time = 106.7440, the second transport starts its sequence while the first is still in motion. The assembly task exits the program once both transports have returned to their ready positions. This sort of multitasking arose naturally from the TranRun4 architecture, but results in an efficient scheduling of the different assembly operations. Additionally, the final assembly sequence is complex but extremely manageable, since it has been built from smaller, thoroughly tested task components.

Chapter 17

THE GLUING CELL EXERCISE IN TRANRUNJ

TranRunJ is a Java implementation of the previously discussed task/state scheduling architecture. In order to illustrate its use, the Glue exercises of chapters 15 and 16 have been rewritten in Java. The following sections illustrate the construction of the same control system with the TranRunJ package.

17.1 Getting Started

The TranRunJ version of the Glue exercise closely parallels the group-priority example in C++. Rather than cover the same ground again, this chapter will use the Glue exercise to illustrate the use of the TranRunJ software. It is assumed that the reader is familiar with the simulation developed in chapter 15.

17.1.1 Program Entry Point

Since Java is a purely object-oriented language, the main entry method must be contained within a class. The standard way to do this in TranRunJ is to create a subclass of `TranRunJ.TrjMain` which includes a static main method for scheduler startup. In addition, an implementation must be provided for the abstract method `userMain()` and several static scheduler properties should be set. An example of this, from the Glue00 exercise, is:

```
public class GlueMain extends TrjMain
{
    /**The application entry point.  Several static properties of the scheduler must
     * be set here.  An instance of the TrjMain class must also be created.  If
     * desired, a custom CTimer can be instantiated and passed in by calling
     * theMain.setTimer().*/
    public static void main(String [] args)
    {
        //  Set the scheduling mode
        //masterSchedulingMode = CMaster.EARLIEST_FIRST_SCHEDULED;
        //masterSchedulingMode = CMaster.SEQUENTIALLY_SCHEDULED;
        masterSchedulingMode = CMaster.MINIMUM_LATENCY_SCHEDULED;
```

```
        // This is currently the only option
        masterThreadingMode = CMaster.SINGLE_THREADED;

        // Choose the timing mode
        masterTimingMode = CMaster.SIMULATED_TIME;
        //masterTimingMode = CMaster.VM_SYSTEM_TIME;
        //masterTimingMode = CMaster.USER_DEFINED;

        // Instantiate the main object
        theMain = new GlueMain ();

        // Set the custom timer, if desired
        //theMain.setTimer(winTimer);

        // Call the startup method and pass in the command line arguments
        trjMain(args);
    }

    /**The scheduler start method is called from userMain.  All Tasks should be
     * instantiated prior to scheduler startup.*/
    protected void userMain ()
    {
        // User code goes here...
    }
}
```

17.1.2 The userMain Method

The body of the overridden `userMain()` method is where all processes and tasks are created. First, however, the scheduler must be provided with an end time and, if running in simulation, an increment for the timer. After this, all necessary processes can be instantiated by creating new `CProcess` objects. Each process object must be given a unique name, which should be a single word since the running processes are specified at the command line. Once the processes are created, all user task objects should be created and each one passed a pointer to the appropriate process object. Any properties which affect all tasks can now be set, such as profiling and state transition auditing. Example syntax is shown next:

```
protected void userMain ()
{
    // Get a reference to the scheduler object
    CMaster theMaster = TrjMain.getMasterScheduler();

    // Set some necessary parameters
    theMaster.setStopTime(2.0); //  in seconds
    theMaster.setTickTime(0.001); //  in seconds

    // Create a Process object to run the Tasks
    process1 = new CProcess ("Process_One");

    // Create the belt Tasks
    beltAxis1 = new BeltAxis (process1, "Belt Axis 1", 5, 0.1, 25.0, 4.0, 2.0);
```

Section 17.2. Writing Custom Tasks and States   317

```
    beltAxis2 = new BeltAxis (process1, "Belt Axis 2", 5, 0.1, 25.0, 4.0, 3.0);

    // A logger Task to record belt data
    beltLogger = new BeltLogger (process1, "Belt Axis logger", 2, 0.1,
    "GlueData00.txt");

    // Turn on some profiling features of the scheduler
    theMaster.doDurationProfile(true);
    theMaster.doTransitionTrace(true);

    // Calling this method starts scheduler execution
    theMaster.go();

    // Any cleanup code can go here...
}
```

In order to start the scheduler, a call must be made to the go() method of the CMaster object.

17.2 Writing Custom Tasks and States

The Task and State base classes in TranRunJ follow a naming convention similar to that of the C++ TranRun4 package. However, creation of user tasks and states in TranRunJ is a bit different. This section describes how to create both types of user clases.

17.2.1 Creating a Task Class

In TranRunJ, the CTask class has been declared **abstract**, so that users must always provide their own subclasses. In addition, there are several **abstract** classes derived from CTask that represent common task types. Creating a custom task class is as simple as subclassing one of these TranRunJ task types. For most applications, this will be either PeriodicTask or ContinuousTask. An example of a PeriodicTask subclass, again from the Glue00 example, is:

```
public class BeltAxis extends PeriodicTask
{
    private double current;          // Current to the motor, amps
    private double inertia;          // System inertia, as viewed from the motor,
                                     // N m s^2

    private double torqueConstant;   // Conversion from motor current to torque,
                                     // N m /amp

    private double position, velocity;

    // References to State objects held by this Task
    private RunState runState = null;

    /**This constructor initializes the physical parameters of the belt, along with
     *  the necessary Task properties.
```

```
 *
 * @param inertia The inertia of the belt axis.
 * @param dampingCoeff The damping coefficient.
 * @param torqueConstant Torque constant used for calculating a simulated
                         velocity.
 *
 * @see TranRunJ.PeriodicTask
 */
public BeltAxis (CProcess proc, String name, int priority, double interval,
                 double inertia, double torqueConstant, double current)
{
    // Call the superclass constructor
    super(proc, name, priority, interval);

    // Adds this Task to the low priority execution list of its parent Process
    addToProcessBackgroundList();

    this.inertia = inertia;
    this.torqueConstant = torqueConstant;
    this.current = current;

    // Create the State object for the single State
    runState = new RunState (this, "Belt Axis Running State");
    setInitialState(runState);

    createGlobalData("Position");
    createGlobalData("Velocity");
}

/**This method is called only once, just prior to scheduler startup.  It's a
 * good place for initialization code.*/
protected void configure ()
{
    position = 0.0;
    velocity = 0.0;
}

/**In this implementation, the belt is always running.*/
public class RunState extends CState
{
    private double lastTime;

    /**A basic State constructor.*/
    protected RunState (CTask task, String name)         {super(task, name);}

    /**This doesn't really need to be here.*/
    protected void entry ()
    {
        lastTime = getTimeNow();
    }

    /**This method performs all the numerical simulation of the belt position.*/
    protected void action ()
    {
        double newTime = getTimeNow();
```

```
            double deltaTime = newTime - lastTime;

            // Do the Euler integration
            double torque = torqueConstant * current;
            velocity += (torque / inertia) * deltaTime;
            position += velocity * deltaTime;

            // Update global data boxes
            copyToGlobalBox("Position", position);
            copyToGlobalBox("Velocity", velocity);

            lastTime = newTime;

            // This is important...all Intermittent Tasks must be idled or they
            // will run again on the next scan
            idle();
        }
    }
}
```

The class definition for the belt axis task may seem simple, but is, of course, built on top of the entire TranRunJ architecture. The `PeriodicTask` class, in particular, allows user-defined code to be executed at regular intervals. Tasks of this type are repeatedly scanned in order of priority and executed after the appropriate interval. Both the priority and the scan interval are set prior to runtime with the class constructor.

Aside from passing parameters to the superclass, the `BeltAxis` constructor is responsible for defining the interaction of this task with others. In this open-loop implementation, no input to the belt axis is required, but information on its state is posted as global data to be recorded by the logger task. In addition, all periodic tasks must add themselves to an execution list prior to scheduler startup. For simulation purposes, instances of `BeltAxis` are placed in the background list by calling `addToProcessBackgroundList()`.

17.2.2 Creating a State Class

Since the executable code of the belt axis task is not that complex, the class could legitimately just override the `taskMain()` method of `CTask` and place the differential equations there. This method would then be called every time the task was scheduled to run. However, in order to illustrate the use of simple state classes, and because the `BeltAxis` class could legitimately become more complex later, an inner class is created to represent the belt axis in a running state. The `RunState` class extends `CState`, and the simulation code is placed in the `action()` method instead. It is not essential for it to be an inner class, however; state classes can be declared as regular top-level classes just like any other.

At this point, it should be noted that the syntax of inner classes in Java has advantages for the relationship between task and state classes. If a state class is declared as an inner class of its parent task, they can freely share private methods and instance variables. For example, all the simulation coefficients as well as the

ODE output are declared as private variables of `BeltAxis`, but are used by the `action()` method of `RunState`. In more complicated task classes with multiple states, this ability rapidly becomes invaluable. In addition, because `CState` is an inner class of `CTask`, any state class automatically has access to all the exposed methods of the `CTask` class.

17.3 Implementing State Transition Logic

In order to create tasks with multiple states, the user will need to specify the conditions that cause the task to transition from one state to another. This is done by providing an overridden version of the `transitionTest()` method of `CState`. If a state transition needs to occur, the method returns a pointer to another state object, or `null` in order to remain in the same state. A representative example is provided by the motion profiler task of Glue03:

```
/**This State accelerates the profiler setpoint until it reaches the cruise
 * velocity.*/
public class AccelState extends CState
{
    //   ...more code here...

    /**Checks to see if the deceleration point has been reached. If so, a
     * transition occurs.  Can also transition to Idle, or also
     * retransition to Accel again.*/
    protected CState transitionTest ()
    {
        int command = -1;

        if (checkForTaskMessage("Command") == true)
        {
            command = readTaskMessageInt("Command");

            if (command == IDLE_COMMAND)
            {
                moveFinished = false;
                return idleState;
            }

            // If a new target comes in during acceleration
            if (command == ACCEL_COMMAND)
            {
                if (checkForTaskMessage("Target") == true)
                {
                    xTarget = readTaskMessageDouble("Target");
                    return accelState;
                }
            }

        }

        // If no new commands have come in, decelerate
        double xStop = calculateStopPoint();
```

```
            if ((direction > 0) && (xStop >= xTarget))
                return decelState;
            if ((direction < 0) && (xStop <= xTarget))
                return decelState;

            return null;
        }
    }
```

There are three transition conditions for the `Accel` state, to either the `Idle` or `Decel` state, or even back to `Accel`. These are all specified by returning a pointer to the appropriate state object. The pointers themselves are members of the enclosing task class, thanks to the inner class capabilities of Java. Note that if a state returns a pointer to itself, as this one does, it effectively causes a retransition into that state. This will result in the `entry()` method being called, just as in a normal state transition.

When developing an application based on multistate tasks, having information about the state transitions is critical to analyzing program flow. Accordingly, a transition audit trail can be recorded for any task by calling the `doTransitionTrace()` method of `CTask`. The state transitions are recorded and collated by the scheduler object, and can be recovered after execution by calling the `writeTransitionTrace()` method of `CMaster`, `CProcess`, or `CTask`.

It is somewhat important to note that state transition auditing will result in memory being allocated for storage at runtime. For any task making very rapid transitions (e.g., a PWM task) it is often a good idea to turn auditing off (with a call to `doTransitionTrace(false)`) once the state transition logic is well-characterized.

17.4 Global Data and Intertask Messaging

In order to facilitate generalized data exchange across processes, TranRunJ includes the same two communication methods discussed in previous chapters. Tasks can either post data to a globally accessible location (a Global Box), or send signals directly to another task via a Task Message. Which mode of communication is appropriate can depend both on the time-critical nature of the data and the number of tasks that are interested in it.

17.4.1 Global Data Items

A typical use of global data exchange is shown by the interaction between the belt axis and belt logger tasks in the Glue00 example. In designing the `BeltAxis` class, it was assumed that other tasks might be interested in the current position or velocity of the belt. By creating two global data items and refreshing them periodically (typically on every execution, but not necessarily), the belt axis task can expose its private data without worrying about who needs it.

The syntax for creating, posting, and reading global data items is simple. To create a global item, the method `createGlobalData()` is used. When called from the task constructor,[1] a new global item with the specified name is created and associated with that task instance. To refresh the data, the `copyToGlobalBox()` method can be called during scheduler execution. The global data can be accessed by other tasks through the `readGlobalDouble()` or `readGlobalInt()` methods.[2] Note that the reading task needs a pointer to the posting task and the name of the box, as in the `BeltLogger` class from Glue00:

```
/**For simple Tasks, the taskMain method can replace a dedicated State class.*/
protected void taskMain ()
{
    // Cast the data into floats so that it displays better
    float time = (float)getTimeNow();

    float belt1pos = (float)readGlobalDouble(GlueMain.beltAxis1, "Position");
    float belt1vel = (float)readGlobalDouble(GlueMain.beltAxis1, "Velocity");

    float belt2pos = (float)readGlobalDouble(GlueMain.beltAxis2, "Position");
    float belt2vel = (float)readGlobalDouble(GlueMain.beltAxis2, "Velocity");

    // ...more code here...
}
```

For small numbers of data items, this is a very low-overhead way of sharing information. Tasks with lots of potential global data may wish to either refresh less critical items infrequently or determine how to send some as task messages.

17.4.2 Task Messages

Task messages basically play the role of e-mail between tasks, whereas global data has more in common with the Web. When one task needs to send data directly to another, the `sendTaskMessage()` method allows one piece of data to be transmitted. The sending task must have a pointer to the receiving task, as well as the name of a message box created by the receiver.

Tasks that are interested in receiving data from others can create inboxes to accept it. Message inboxes are instantiated by calling `createMessageInbox()` from the task constructor. Retrieving incoming data from these is a two-step process, involving first testing for the presence of new data, followed by collection if information is present. For example, the `PIDController` class from Glue02 checks several inboxes on each execution:

```
/**This method does all the work for this task.  Calculates a PID actuation signal
 * based on the current difference between the setpoint and process value.*/
protected void action ()
{
    double deltatime = getTimeNow () - lastTime;
```

---
[1] Both global data and task message boxes can be created in state constructors as well; these are automatically associated with the parent task.

[2] For simplicity, both global data and task messages only handle double or integer data. Internally, the value is stored as a double.

```
    //  This is a lot of message boxes to check...
    //  there's got to be a better way of doing this.
    if (checkForTaskMessage("Setpoint") == true)
        setpoint = readTaskMessageDouble("Setpoint");
    if (checkForTaskMessage("Process Value") == true)
        processValue = readTaskMessageDouble("Process Value");

    if (checkForTaskMessage("MaxActuation") == true)
    {
        maxActuation = readTaskMessageDouble("MaxActuation");
        if (clipOutput == false)  clipOutput = true;
    }

    if (checkForTaskMessage("MinActuation") == true)
    {
        minActuation = readTaskMessageDouble("MinActuation");
        if (clipOutput == false)  clipOutput = true;
    }

    if (checkForTaskMessage("Kp") == true)
        Kp = readTaskMessageDouble("Kp");
    if (checkForTaskMessage("Ki") == true)
        Ki = readTaskMessageDouble("Ki");
    if (checkForTaskMessage("Kd") == true)
        Kd = readTaskMessageDouble("Kd");

    //  More code here...

    //  Update the global value and the listening Task, if any
    copyToGlobalBox("Actuation", actuation);

    if (controlTask != null)
        sendTaskMessage(controlTask, controlBox, actuation);

}
```

This setup allows the PID task class to be written without any knowledge of what signal it is calculating a control value for. In the Glue02 example, the belt axis task sends its position to the Process Value inbox of the PID task as the control variable and reads actuation back from global data. However, the PID controller can also be connected directly to a single task that needs to be controlled. This messaging scheme, then, provides a convenient mechanism for tasks to expose inputs that others can use. The only caveat is that boxes must be checked as often as possible to ensure data is not missed or overwritten, at least in the current implementation.

17.5 Continuous vs. Intermittent Tasks

The Glue04 exercise introduces the use of the continuous task type, as opposed to the intermittent (or more specifically, periodic) tasks used previously. The Transport task classes are intended to be an intermediate supervisory layer in the final control scheme, and are responsible for coordinating the conveyor belts, retaining clamps, and unloading robots. There are no numerical calculations associated with these tasks that need to be run at particular times, and the signals they send would probably not be very time-critical in an actual system. The `Transport` tasks just move through a predefined sequence of states, generating appropriate messages for the other tasks they supervise.

These looser constraints make `Transport` a good candidate for extending the `ContinuousTask` class. Continuous tasks are the lowest priority type in TranRunJ, and are all run in round-robin fashion when no intermittent tasks need to. Depending on how many there are, and how heavily the scheduler is loaded with higher-priority intermittent tasks, continuous tasks may get quite a bit of processor time. However, when it is time for an intermittent task to run, any continuous task will have to wait. In this case, as long as there aren't a lot of motion profilers and PID control tasks running at the same time, the signals sent by the `Transport` task should be sent promptly. If this were ever a problem, the matter could probably be solved just by making `Transport` a `PerodicTask` as well.

At this point, the `IntermittentTask` class deserves a bit of explanation. There are two intermittent task subclasses in the TranRunJ package: `PeriodicTask` and `EventTask`. Periodic tasks were introduced early in the Glue examples, since they are well suited for tasks that require scheduler time on a defined interval. Event tasks obey the same priority rules as periodic tasks, but become runnable in response to an event of some kind. This could be any of several things: expiration of a separate timer, arrival of a message from another task, or even a signal from external hardware. The user-defined `checkEvent()` method of any event task is constantly tested to see if it evaluates to `true`. If so, the event task will run at the next opportunity, subject to its priority level. Event tasks can be useful for high-priority operations that do not require frequent polling of user action code or transition conditions.

17.6 Scheduler Internals

Since TranRunJ is a singly threaded, cooperatively scheduled runtime environment, the scheduler is, in essense, just one big `while` loop. Tasks execute continuously, at specified times or in response to events. However, understanding the internal organization of the scheduler can help in getting the most out of a TranRunJ application.

17.6.1 Operating System Processes vs. CProcess

The process classes in TranRunJ are more analogous to system processes than tasks are to threads. All tasks must be added to a process prior to scheduler startup, and the process class separates the tasks into execution lists of appropriate priority. Typically, only one running process instance would be created per virtual machine, and hence per processor. All other process objects are instantiated, but the tasks they hold are not executed. However, for debugging or simulation purposes, more than one process can be run in the same scheduler. In this case, processes get the chance to run their tasks in a simple round-robin fashion. Only in the case where a single CProcess is instantiated per scheduler does it correspond directly to a running OS process.

17.6.2 Foreground vs. Background Execution Lists

The terms foreground and background may be somewhat misleading here, since they are sometimes used to refer to tasks executing in interrupt service routines. In the case of TranRunJ, the difference between the two is simpler. Each CProcess object has these two execution lists for intermittent tasks, along with a separate one for continuous tasks. Tasks in the foreground list are considered higher priority and are given the chance to run before any potential interprocess communication takes place. Background intermittent tasks are allowed to execute once any relevant data is exchanged with other processes running elsewhere.

17.6.3 Scheduling Modes

CMaster and CProcess currently support three scheduling modes: sequential, minimum-latency, and earliest-deadline-first. Sequential scheduling involves polling all tasks in the order of instantiation (specified at compile-time) while ignoring relative priority. In earliest-deadline, the scheduler continually determines the next execution time of all intermittent tasks, finally calling the schedule() method of whatever task should run next. In the meantime, continuous tasks are allowed to run. The minimum-latency mode involves the most computational overhead, but allows the priority structure to be used. Intermittent tasks are given the chance to run in priority order, with continuous tasks being allowed to run in the meantime.

17.7 Execution Profiling

When developing an efficient control system, it is critical that the application be able to examine itself so that performance can be evaluated. The TranRunJ.Profile subpackage contains a few simple classes that allow task behavior to be measured at runtime. The CProfiler class is available for just this reason, and its use has been incorporated into nearly every aspect of the TranRunJ package. In fact, a good deal of the scheduler, task, and state code contains provisions for conditional execution profiling. All of the core TranRunJ classes expose methods for setting different profiling options, including duration of task execution, task latencies, and a number of scheduler properties. The CProfiler class can collect successive time difference measurements and perform some basic statistics on these. In addition, some profiling methods of CTask are declared protected, so that user subclasses can override them for custom profiling.

A few key methods allow the user to collect quite a bit of information about task durations and latencies. Durations, in this case, reflect the time it takes all user code for a task to run. This typically includes the action() and transitionTest()

methods, and also the `entry()` method on those scans where it is called. Latencies are just the difference between the scheduled and actual execution times of intermittent tasks. The `doDurationProfile()` method is available for any of the `CMaster`, `CProcess`, and `CTask` classes.

The information collected by these methods can be important in determining performance bottlenecks, as well as for optimizing relative task priorities. Data for any task can be collected in histogram form by calling `doDurationHistogram()` and similar. Finally, the current state of the scheduler can be captured at any time by calling the `writeTaskStatus()` method of `CMaster`.

17.8 Intertask Messaging Across Different Processes

One fundamental aspect of the TranRunJ architecture is that the specifics of intertask messaging are hidden from the user as much as possible. In principle, this allows tasks to communicate seamlessly, regardless of where they are running. Tasks in the same OS process, different OS processes, or on another machine entirely should all be able to use the same syntax to send each other data. Enforcing this architecture results in a great deal of flexibility when implementing intertask messaging between different processes.

In the current version of TranRunJ, all intertask messaging (except when it occurs between two tasks with the same `CProcess` parent) is transmitted using a UDP protocol. The specific code is introduced by subclassing `CProcess` and overriding a number of methods. Of course, different communication protocols could easily be used, just by creating different subclasses. The structure of the messages themselves is quite simple, with only a few pieces of data needed for directions. The entire specification for the UDP implementation, both for global data and task messages, can be found in the `IpProcess` source code.

Creating processes on separate computers is straightforward: simply create instances of `IpProcess` where `CProcess` was used before. Then add tasks to the appropriate `IpProcess` objects. For example, in the Glue01 exercise, two sets of belt axis and oven tasks are instantiated as follows:

```
protected void userMain ()
{
    //  ...more code here...

    // Create a Process object to run the Tasks
    process1 = new CProcess ("Process_One");

    // Create the belt Tasks
    beltAxis1 = new BeltAxis (process1, "Belt Axis 1", 5, 0.1,
                              25.0, 4.0, 3.0);

    beltAxis2 = new BeltAxis (process1, "Belt Axis 2", 5, 0.1,
                              25.0, 4.0, 2.0);

    // Create the oven Tasks
    thermalOven1 = new Oven (process1, "Oven 1", 5, 0.1,
                             0.0, 20.0, 10.0, 20.0, 2e5);
```

Section 17.8. Intertask Messaging Across Different Processes

```
        thermalOven2 = new Oven (process1, "Oven 2", 5, 0.1,
                                 0.0, 20.0, 7.0, 20.0, 2e5);

    //  ...more code here...
}
```

Since the example is intended to run on a single computer, only one CProcess object is created. By passing a pointer to the process through their constructors, each task is added to a process execution list. When this program is run from the command line, the name of the process is specified so that the scheduler will know it is intended to run locally. If, instead, each belt axis and oven task pair were supposed to run on separate computers, the code might look like this:

```
protected void userMain ()
{
    //  ...more code here...

    //  Create a Process object to run the Tasks
    process1 = new IpProcess ("Process_One", "192.168.1.100");
    process2 = new IpProcess ("Process_Two", "192.168.1.101");

    //  Create the belt Tasks
    beltAxis1 = new BeltAxis (process1, "Belt Axis 1", 5, 0.1,
                              25.0, 4.0, 3.0);

    beltAxis2 = new BeltAxis (process2, "Belt Axis 2", 5, 0.1,
                              25.0, 4.0, 2.0);

    //  Create the oven Tasks
    thermalOven1 = new Oven (process1, "Oven 1", 5, 0.1,
                             0.0, 20.0, 10.0, 20.0, 2e5);

    thermalOven2 = new Oven (process2, "Oven 2", 5, 0.1,
                             0.0, 20.0, 7.0, 20.0, 2e5);

    //  ...more code here...
}
```

Initially, this setup can be somewhat misleading. Although both processes and all four tasks are instantiated locally (that is, distinct instances of each exist on both target machines simultaneously), they will not necessarily run locally. That is determined by which processes the user specifies to be run when the program is started.

In this example, there will be two processes running on different machines, each with a separate IP address. On the machine with the IP address of 192.168.1.100, the running process will be "Process_One" and its running tasks will be "Belt Axis 1" and "Oven 1." The opposite will be true for the other machine. The key here is that both copies of the scheduler program create instances of every task that will run anywhere. Tasks belonging to processes that are not running locally will never have their state methods called. Rather, they act as placeholders for their counterparts running elsewhere, holding global data and task message inboxes that running tasks can access.

Although one clear drawback of this scheme is increased memory and space requirements, those considerations should be outweighed in many cases by the increased ease of testing and flexibility in distributing tasks between processors. The scheduler is not limited to running a single process; if simulated multiprocess operation is desired, several running processes can be specified at the command line. The ability to simulate the entire system on single machine is an excellent way to refine program flow and eliminate bugs at an early stage. Any task can later be assigned to any of the processes, and will be executed only in the scheduler which is running that process.

17.9 Tips And Tricks

As with anything, there are a number of small efficiencies that can be squeezed from TranRunJ by those with the patience to do so. The following represent some changes that can actually have a measurable performance difference when applied together.

17.9.1 Judicious Use of Execution-Time Profiling

When all of the profiling features of TranRunJ are enabled simultaneously, there is quite a bit of data being collected all the time. With larger numbers of tasks, this can have a serious impact on performance. Obviously, all the collected data must be placed somewhere, and in Java this typically means memory allocation events at runtime. Execution and latency profiling are off by default, and these should really only be used by a few tasks at a time. Of course, it is generally a good idea to avoid memory allocation at runtime whenever possible.

17.9.2 Integer Labels for Global Data and Task Message Inboxes

Another noticeable performance bottleneck is the use of objects for global data and task message inbox identification. While this is a nice feature for students, in practice it can add tens of microseconds per box to state method execution. A specific example of this is the PIDController task, which must check for five incoming task messages on every execution of the action() method. However, both types of box objects are also assigned integer serial numbers upon instantiation. These simply start from zero with the first box of either type. CTask has methods for accessing global data and task message inboxes by number which are similar to those which use strings.

17.9.3 The TaskMessageListener Interface

Another marginal performance gain can be had by reducing the number of times each message inbox must be scanned for the arrival of new data. If, instead, a

Section 17.9. Tips And Tricks

TaskMessageListener object is registered with each box, the data can instead be pushed in. The following is an example of this from the Assembly class in Glue08:

```
createMessageInbox(BASE_DONE);

addTaskMessageListener(BASE_DONE, new TaskMessageListener () {
    public void messageReceived (CTask toTask, String name, double value, CTask fromTask) {
        if ((int)value == LoadRobot.LOAD_READY)
            baseDone = true;
    }
});
```

While the syntax here may seem odd initially, it is reasonably straightforward. TaskMessageListener is an interface,[3] which means instances cannot be directly created. Instead, an anonymous class which *implements* the interface is instantiated and passed directly to the addTaskMessageListener() method. In this case, the anonymous class is overriding the messageReceived() method of the interface with code appropriate to that inbox.

The more typical implementation of this code would have the task polling the inbox on every scan to check for new data. Instead, the baseDone variable can be updated instantly when the messageReceived() method is called from inside the message box. The only critical issue here is for code in these callback methods to remain very short, since TranRunJ is singly threaded and this method must be called during the *sending* task's execution scan.

17.9.4 Scheduler Sleeping

Ideally, the TranRunJ scheduler is intended to run in a real-time environment as the only process. However, one of the advantages of Java and the TRJ architecture is that the majority of a control program can be simulated on simple workstations prior to real-time execution. This can, unfortunately, lead to the control program taking most of the processor time, especially if any native libraries are loaded. In order to make testing on regular OS platforms easier, the scheduler has the ability to sleep its thread when no intermittent tasks need to be run. Calling the doSleepWhenPossible() method of CMaster prior to scheduler startup will do this. The one caveat (as there is always one) is that this feature does not coexist well with continuous tasks, since by definition they are supposed to be running when no intermittent tasks are. The current version of TranRunJ does not support using this feature and continuous tasks simultaneously.

17.9.5 Anonymous State Classes

While not really a performance issue, sometimes the creation of lots of state classes can be a chore. If a state has short executable code and does not define any

---

[3]Interfaces are Java's replacement for the multiple inheritance features of C++. See any Java text for a more thorough discussion.

methods beyond those it inherits, it can easily be created as an anonymous class. For example, the Start state from Glue08.Assembly:

```
public class Assembly extends ContinuousTask
{
    // References to State objects held by this Task.  These can all be declared
    // as type CState, since none of them define methods of their own.
    private CState startState, beginNewAssemblyState,
                placeBaseState, applyGlueState,
                placeCylinderState, waitForTransportState,
                cureUnloadState, stopState;

    // ...more code here...

    public Assembly (CProcess proc, String name, int numToMake)
    {
        // ...more code here...

        startState = new CState (this, "Start") {
            protected CState transitionTest () {
                // If both transports are ready, start a new assembly
                if ((readGlobalInt(GlueMain.transport1, "Transport Status") ==
                            Transport.TRANSPORT_READY
                    && (readGlobalInt(GlueMain.transport2, "Transport Status") ==
                            Transport.TRANSPORT_READY))
                    return beginNewAssemblyState;

                return null;
            }
        };

        // ...more code here...
    }

    // ...more code here...
}
```

In this case, the Start state only overrides the transitionTest() method of CState, since it is only meant to provide a well-defined starting point for the Assembly task. This method has inner class access rights to the beginNewAssemblyState pointer, and hence can return it when the state transition is made. Similarly, the startState pointer is of type CState, is perfectly valid, and could be used by any other state object. This simple class definition is possible because the CState class handles almost all the specifics of state code execution, as well as registering the state object with its parent task. While not suitable for every situation, anonymous classes such as this offer a compact way to define simple states.

17.10   Additional Information

Other TranRunJ resources, such as updated versions, additional documentation, and a FAQ document can be found at http://www.ugcs.caltech.edu/~joeringg/TranRunJ.

BIBLIOGRAPHY

[1] Lindsey Vereen, "Vehicular Rant," *Embedded Systems Programming*, p. 5, April 2000.

[2] Scott Briggs, "Manage your Embedded Project," *Embedded Systems Programming*, pp. 26–46, April 2000.

[3] Nancy G. Leveson and C. S. Turner, "An Investigation of the Therac-25 Accidents," *IEEE Computer*, pp. 18–41, July 1993.

[4] Jack G. Gansle, "Embedded Y2K," *Embedded Systems Programming*, p. 97, February 1999.

[5] Paul T. Ward and S. J. Mellor, *Structured Development for Real-Time Systems*. Prentice-Hall, 1985.

[6] David M. Auslander, A. C. Huang, and M. Lemkin, "A Design and Implementation Methodology for Real Time Control of Mechanical Systems," *Mechatronics*, vol. 5, no. 7, pp. 811–832, 1995.

[7] David M. Auslander, "Unified Real Time Task Notation for Mechanical System Control," American Society of Mechanical Engineers, 1993.

[8] David M. Auslander, M. Lemkin, and A. C. Huang, "Control of Complex Mechanical Systems," in *Proceedings of 1993 IFAC Congress*, International Federation of Automatic Control, 1993.

[9] Zvi Kohavi, *Switching and Finite Automata Theory*. McGraw-Hill, 1970.

[10] Richard S. Sandige, *Modern Digital Design*. McGraw-Hill, 1990.

[11] David A. Dornfeld, D. M. Auslander, and P. Sagues, "Programming and Optimization of Multi-Microprocessor Controlled Manufacturing Processes," *Mechanical Engineering*, vol. 102, no. 13, pp. 34–41, 1980.

[12] Koen Bastiaens and J. Van Campenhout, "A Visual Real Time Programming Language," *Journal of Control Engineering Practice*, no. 1, pp. 59–63, 1993.

[13] Paul Le Guernic, A. Benveniste, P. Bournai, and T. Gautier, "Signal, A Data-Flow Oriented Language for Signal Processing," *IEEE Trans. on Acoustics, Speech and Signal Processing*, no. 34, pp. 362–374, 1986.

[14] Albert Benveniste and P. Le Guernic, "Hybrid Dynamical Systems Theory and the Signal Language," *IEEE Trans. on Automatic Control*, vol. 35, no. 5, pp. 535–546, 1990.

[15] David M. Auslander and C. H. Tham, *Real-Time Software for Control: Program Examples in C*. Prentice-Hall, 1990.

[16] Kris Jamsa and K. Cope, *Internet Programming*. Jamsa, 1995.

[17] Pat Bonner, *Network Programming with Windows Sockets*. Prentice-Hall, 1995.

[18] Bill Joy, G. Steele, J. Gosling, and G. Bracha, *The Java Language Specification, Second Edition*. Addison-Wesley, 2000.

[19] Scott Oaks and H. Wong, *Java Threads, Second Edition*. O'Reilly and Associates, 1999.

[20] Allen Holub, *Programming Java Threads in the Real World*. 1998. www.javaworld.com.

[21] David Flanagan, *Java In A Nutshell, Second Edition*. O'Reilly and Associates, 1997.

[22] Greg Bollella, B. Brosgol, P. Dibble, S. Furr, J. Gosling, D. Hardin, and M. Turnbull, *The Real-Time Specification for Java*. Addison-Wesley, 2000.

[23] J Consortium, "International j consortium specification real-time core extensions," 2000. www.j-consortium.org.

[24] George L. Batten, *Programmable Controllers, Hardware, Software & Applications*. McGraw-Hill, 1994.

[25] Gilles Michel, *Programmable Logic Controllers, Architecture and Applications*. John Wiley & Sons, 1990.

[26] Ian G. Warnock, *Programmable Controllers, Operation and Application*. Prentice-Hall, 1988.

[27] J. Holmes, "State Machine Techniques Develop PLC Ladders," *Industrial Computing*, pp. 12–15, April 1996.

INDEX

abstract classes, 317
AcceptDataPack() function, 115
AcceptMessage() function, 110
action function, *see* state functions
Action() function, 119, 279
action() method, 319
actuator, 3, 22
AddProcess() function, 287
addTaskMessageListener() method, 329
amplifier, 4
    switching vs. linear, 38
applets, 217
application layer, *see* networking
assembly language, 46
Assembly task, 267, 309, 329
asynchronous
    operations, 8
    threads, 20
audit trail, *see* transition audit trail

background vs. foreground, 84
BaseTask class, 68, 108, 110, 116, 240
    AcceptMessage() function, 110
    constructor, 77
    CopyGlobal() function, 116, 117
    CreateGlobal() function, 116
    GetGlobalData() function, 116, 117
    GetMessage() function, 109
    RegisterStateName() function, 261
    Run() function, 68
    SendMessage() function, 108, 110, 249, 260
belt axis, *see* conveyor, 275

blocking code, 41
Bridgeview, 162
    and DDE, 180
    virtual instrument, 164
buses, 186

C language, 46
C++ language, 46
    simulation in TranRun4, 82
    templates for simulation, 68
callback functions, 156
cam grinder, 6
CDataLogger class, 280
cell phone, 6
checkEvent() method, 324
CheckForDataPack() function, 114
checkForTaskMessage() method, 321, 323
CMaster class
    in TranRun4, 84, 287
        AddProcess() function, 287
        Go() function, 84, 288
    in TranRunJ, 325
        go() method, 317
        writeTransitionTrace() method, 321
communication
    between computers, 12
    between processes, 105
    between tasks, 22, 61, 99
    in TranRun4, 284
    in TranRunJ, 322
    client-server, 173, 196

media, 105
peer-to-peer, 173
push vs. pull, 99, 284
within a process, 100
computational technology, 9
continuous tasks, 88
software realization of, 317
vs. intermittent tasks, 128
ContinuousTask class, 317, 324
control
definition, 5
distributed, 217
embedded, 6
feedback, 5, 23
PID for gluing cell, 291
position, 27
temperature, 24
thermal systems, 24
velocity, 23
conveyor, 275
dynamic model, 275
task, 277
cooperative multitasking
in Windows, 62
COperatorWindow class, 149, 294
AddInputItem() function, 150
AddKey() function, 150, 295
AddOutputItem() function, 150
Close() function, 150
constructor, 149
Display() function, 150, 296
Update() function, 150, 296
CopyGlobal() function, 116, 117
CopyGlobalData() function, 94, 119, 280, 286
copyToGlobalBox() method, 322, 323
CProcess class
in TranRun4
WriteConfiguration() function, 289
WriteStatus() function, 290
in TranRunJ, 316, 324, 326
writeTransitionTrace() method, 321
vs. operating system process, 324

CreateDataInbox() function, 114, 285
CreateGlobal() function, 116
CreateGlobalData() function, 93, 119, 279, 286, 304
createGlobalData() method, 322
createMessageInbox() method, 322, 329
critical section, 102, 117
CriticalSectionBegin() function, 80, 102, 117, 245
CriticalSectionEnd() function, 80, 102, 117, 245
CState class
in TranRun4, 82, 277
Action() function, 119, 279
Entry() function, 119, 279
Idle() function, 280
TransitionTest() function, 119, 279, 283
in TranRunJ
and anonymous classes, 330
CTask class
constructor in TranRun4, 304
in TranRun4, 82, 113, 277, 288
AcceptDataPack() function, 115
CheckForDataPack() function, 114
CopyGlobalData() function, 94, 119, 280, 286
CreateDataInbox() function, 114, 285
CreateGlobalData() function, 93, 119, 279, 286, 304
ReadDataPack() function, 114, 280, 286, 302
ReadGlobalData() function, 119, 286, 292
Run() function, 119
SendDataPack() function, 114, 285, 292
SetInitialState() function, 91, 304
in TranRunJ, 317
addTaskMessageListener() method, 329

checkForTaskMessage() method, 321, 323
copyToGlobalBox() method, 322, 323
createGlobalData() method, 322
createMessageInbox() method, 322, 329
doTransitionTrace() method, 321
readGlobalDouble() method, 322
readGlobalInt() method, 322
readTaskMessageInt() method, 321, 323
sendTaskMessage() method, 322
taskMain() method, 319
writeTransitionTrace() method, 321
inTranRun4
SetInitialState() function, 87

data
    exchange by function call, 103
    storage of task's, 302
data integrity, 100
    design rules, 102
data logger
    in GPP scheduler, 239
    in TranRun4, 280
data ownership, 115
data pack, 112, 284
debugging
    tools in TranRun4, 289
design
    "bottom-up" approach, 274
    documentation, 13
    hierarchical, 21
    principles, 8
dispatcher, see scheduler
distributed control, 207, 217
distributed database, 115
    in GPP scheduler, 116, 248
    in the TranRun4 scheduler, 118, 286
    in the TranRunJ scheduler, 321
    single process, 117

documentation
    design, 13
    with javadoc, 219
doTransitionTrace() method, 321
dynamic model
    oven, 288

economic tradeoffs, 16
embedded computers, 1
entry function, see state functions
Entry() function, 119, 279
entry() method, 321
Euler integrator, 51, 237, 274, 276
EventTask class, 324
    checkEvent() method, 324
exchange variables, 61, 103
execution loop
    in Matlab template, 51
    in TranRun4, 83
    in TranRunJ, 324
exit function, 42
extern variables, 104
EXTERNAL definition, 76

feedback control, 5
    in GPP Glue example, 247
fly-by-wire control, 11
foreground vs. background, 84

G language, 162, see Bridgeview
GetGlobalData() function, 116, 117
GetMessage() function, 109
global data, see distributed database
global database, see distributed database
global variables, 61, 104, 303
gluing cell, 235, 273, 315
Go() function, 84, 288
go() method, 317
graphical operator interface
    see GUI, 155
Group Priority (GPP) Scheduler, 67, 106, 116
    list classes, 70
    vs. TranRun4, 82

GUI (graphical user interface), 12, 155
    callback functions, 287
    in Java, 214

heater, 24

Idle() function, 280
idle() method, 319
inbox, *see* message inbox, 285
inheritance, 137, 218
inner classes, 218, 319, 330
Intel 8254 chip, 123
intermittent tasks, 88
    software realization of, 317
    vs. continuous tasks, 128
IntermittentTask class, 324
    idle() method, 319
interrupt service routine (ISR), 125, 142
interrupts, 44, 61, 138
    sources, 141
    using, 143
IP (Internet Protocol), 190
IpProcess class, 326

Java language, 46, 213
    abstract keyword, 317
    anonymous classes, 329
    applets, 217
    garbage collection in, 215
    hardware access in, 215
    HTML documentation in, 219
    inner classes in, 218
    networking in, 214, 218
    tasks and states in, 218
Java Virtual Machine (JVM), 215
    bytecodes for, 217
java.net, 214, 218
javadoc, 219
javax.comm, 216

ladder logic, 221, 224
LAN (local area network), 155
line shaft, 31

main() function, 287
master scheduler, 84, 287, 325
Matlab, 46
    command-line, 157
    for audit trail analysis, 262
    for control systems, 46
    minimum latency dispatcher, 131
    PLC simulation with, 232
    templates for simulation, 48
mechatronics
    definition, 1
message box, *see* message inbox
message inbox, 112
    in GPP scheduler, 107
    in TranRun4, 112, 285
    in TranRunJ, 322
message passing, 105
    in GPP scheduler, 106, 248
    in the TranRun4 scheduler, 112
    in the TranRunJ scheduler, 322
    multiple processes, 209, 326
    single process, 110
MessageBox class, 107
minimum latency scheduler
    in TranRunJ, 325
    template in Matlab, 131
motion
    coordinated, 31
    profile, 31, 59, 299
motor
    DC, 23, 27, 275
MS-DOS
    command-line, 147
    hardware timers, 123
    prompt, 246
    screen output latencies, 294
    Timer 0, 125
    timing, 97
    windowing programs in, 157
multitasking, 51
    cooperative, 43, 133, 324
    preemptive, 138
    with PLCs, 230
mutual exclusion, 102

networking, 187
    application layer, 197, 203
    in Java, 218
    non-blocking, 195, 211
    packets, 203
    transport layer, 191
    Winsock model, 193, 208
    with TCP, 190, 191
    with UDP, 191
new operator, 71
non-blocking code, 41, 221

ODE solver, 51, 274, 276
one-shot-rising (OSR) element, 225
operator interface, 11
    and non-blocking code, 158, 294
    character-based, 145
    context sensitive, 146
    design rules, 147
    for glue project, 292
    for heater program, 151
    in Java, 217
    timing, 154
optical encoder, 22, 27
oven control, 26, 288

packet, *see* networking
PDA (personal digital assistant), 6
performance specification, 12
performance timer, 124
PeriodicTask class, 317, 324
PID control, 291
PLC (programmable logic controller), 15, 35, 221
    simulation with, 222
port
    network, 194
portability, 12, 46
preemption, 61, 102, 139
preprocessor macros, 107, 239
priority, 141
    based scheduling, 127
process, 10, 20
    vs. CProcess objects, 324

prototype
    laboratory, 15
    production, 16
PWM (pulse-width modulation), 38
    simulation in C++, 80
    simulation in Matlab, 53

quality control, 13

ReadDataPack() function, 114, 280, 286, 302
ReadGlobalData() function, 119, 286, 292
readGlobalDouble() method, 322
readGlobalInt() method, 322
readTaskMessageInt() method, 321, 323
real-time operating system, 44
real-time software, 5
    from simulation, 62
    in C++, 97
    in Java, 215
RegisterStateName() function, 261
relay logic, 222
rollover
    of MS-DOS hardware timer, 126
    of real-time clock, 122
Run() function
    in GPP scheduler, 68
    in TranRun4, 119

safety, 32
scanf() function, 41
scheduler
    cooperative, 129
    earliest deadline first, 325
    group priority, 67
    interrupt-based, 140, 142
    minimum latency, 130, 325
    sequential, 325
    TranRun4, 67, 273
SendDataPack() function, 114, 285, 292
SendMessage() function, 108, 110, 249, 260
sendTaskMessage() method, 322

sensor, 22
sequential logic
    asynchronous, 223
    synchronous, 223
SetInitialState() function, 87, 91, 304
shared global data, *see* distributed database
shared memory, 105, 187, 219, 284
Signal language, 35
simulation, 14, 29–30, 48
    and PLCs, 222
    heater with PWM, 131
    in C and C++, 67
    in C++, 82
    in Java, 217
    oven, 288
    running in real time, 62
SOAP, 217
socket, 214
sockets, 192
software
    conventional vs. real-time, 5
    maintenance, 17
    modular, 9
    nasty properties, 8
    portability, 10
    real-time, 5
state, 10, 19
    functions (entry, action, and test), 41
    in TranRun4, 277
    naming convention, 277
    scan, 42, 51
    transition, 36
state transition logic, 35
    and PLCs, 227
symmetric multiprocessing, 186
system
    components, 3
    design principles, 8
    engineering, 9
    multi-computer, 12
    multi-processor, 99

task, 10, 19
    classes vs. objects, 75

design
    examples, 22–29, 31–33
    diagrams, 22
    dispatcher, *see* scheduler
    in GPP Scheduler, 68
    in TranRun4, 277
    in TranRunJ, 317
    lists, 134, 325
    priority, 20
    storage of data, 302
    supervisory, 58
task messages, *see* message passing
TaskDispatcher() function, 74, 133
taskMain() method, 319
TaskMessageListener interface, 328
TCP (Transport Control Protocol), 173, 190, 191
    in Java, 214
TCP/IP, *see* TCP
templates
    C++
        simulation, 68
        task, 77
    Matlab
        simulation, 48
        task, 51
testing, 13
Therac-25 (radiation therapy machine), 7
thread, 10, 20
    Java Thread class, 214
three-tank system, 21, 39, 57
tic & toc in Matlab, 63
TICKRATE definition, 97, 135, 245
time, 47
    calibrated, 47, 62, 121
    measured, 62, 122
    measuring in C++, 97
    measuring in Java, 216
    measuring in Windows, 124
time-slice, 140
TIMEINTERNAL definition, 135, 245
timer, 27
    free-running, 97, 122
    hardware, 123

in Matlab, 63
internal, 97
interrupt, 125, 138
overflow, 122
timing mode
in GPP scheduler, 245
TList class, 70
TranRun4 scheduler, 67
TranRunJ scheduler, 218, 315
transition, 36
audit trail, 55, 305
trace, *see* audit trail
transition logic, 35
diagram, 36
in TranRunJ, 320
table, 37, 59
transition test function, *see* state functions
TransitionTest() function, 119, 279, 283
transitionTest() method, 320
transport layer, *see* networking
Transport task, 260, 307, 309, 323
TrjMain class, 315
userMain() method, 315

UDP (User Datagram Protocol), 191, 203
assembling packets, 205
in Java, 214
unit machine, 4
universal real-time solution, 43
UNIX
networking, 191
timing in, 124
UserMain() function, 94, 287
userMain() method, 315

virtual functions, 68
virtual machine, *see* Java Virtual Machine
Visual Basic, 160

washing machine, 32
Watt governor, 2

Windows
and Visual Basic, 160
DDE communication in, 173
multitasking, 62
networking, 193
thread priorities, 215
timing, 97
timing in, 124
user interface programming, 147
Winsock model, 193, 208
writeTransitionTrace() method, 321

XML, 217

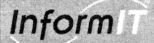  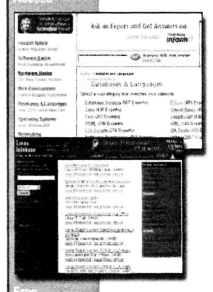

PRENTICE HALL
Professional Technical Reference
Tomorrow's Solutions for Today's Professionals.

Keep Up-to-Date with
PH PTR Online!

We strive to stay on the cutting edge of what's happening in professional computer science and engineering. Here's a bit of what you'll find when you stop by **www.phptr.com**:

- **Special interest areas** offering our latest books, book series, software, features of the month, related links and other useful information to help you get the job done.

- **Deals, deals, deals!** Come to our promotions section for the latest bargains offered to you exclusively from our retailers.

- **Need to find a bookstore?** Chances are, there's a bookseller near you that carries a broad selection of PTR titles. Locate a Magnet bookstore near you at www.phptr.com.

- **What's new at PH PTR?** We don't just publish books for the professional community, we're a part of it. Check out our convention schedule, join an author chat, get the latest reviews and press releases on topics of interest to you.

- **Subscribe today! Join PH PTR's monthly email newsletter!**

  Want to be kept up-to-date on your area of interest? Choose a targeted category on our website, and we'll keep you informed of the latest PH PTR products, author events, reviews and conferences in your interest area.

  Visit our mailroom to subscribe today! **http://www.phptr.com/mail_lists**